ENGINEERING

The Way I See It

GERALD W. MAYES, PE, RETIRED

Copyright © 2023 Gerald W. Mayes, PE, Retired
All rights reserved
First Edition

Fulton Books
Meadville, PA

Published by Fulton Books 2023

ISBN 979-8-88731-180-7 (paperback)
ISBN 979-8-88731-181-4 (digital)

Printed in the United States of America

CONTENTS

Preface v
Acknowledgments vii

Chapter 1:   Background 1
Chapter 2:   Training 6
Chapter 3:   Formal Education 11
Chapter 4:   A Summer Job 23
Chapter 5:   Return to School and More Education 34
Chapter 6:   My First Real Job—What Does a Civil
             Engineer Do? 53
Chapter 7:   On-the-Job Training 70
Chapter 8:   In-House Training versus Actual Performance ... 78
Chapter 9:   Handling Field Problems 83
Chapter 10: Sent to the Field—On My Own 97
Chapter 11: My Inexperience Shows 107
Chapter 12: Gaining Experience on Additional Field
             Assignments 113
Chapter 13: Promoted to First Line of Supervision 122
Chapter 14: Additional Duties as CADD Manager 133
Chapter 15: Transferred to Construction—the Good, the
             Bad, and the Ugly 142
Chapter 16: Back to Design 167
Chapter 17: Owner's Engineering to Contract Engineering 176
Chapter 18: Additional Experience as a Critical Lift Specialist ... 187
Chapter 19: Consulting for the Construction Group 193
Chapter 20: Modular Design Experience 204
Chapter 21: Nuclear Design Support Experience 211

Chapter 22: Management Experience 216
Chapter 23: Training Experience.. 222
Chapter 24: Proper Engineering Projects 227
Chapter 25: Thoughts on the Engineering Profession 245
Chapter 26: Postretirement Experience 249
Chapter 27: A Look Back on My Career 255

Appendices 259

PREFACE

This is a book about my life experiences related to the field of engineering, specifically civil engineering. It does not include all my experiences but rather gives examples and instances that helped shape my career and directions in life. It is drawn from my recollections and memories that affected me the most and relays both good and bad experiences. The good experiences helped me to build on my knowledge and reinforce good habits and practices. The bad experiences helped me to redirect my efforts and learn good lessons on what not to do. I have been helped and supported along the way by the many people I have worked for, worked with, and those who worked for me. My experiences and skills have been learned and used for the benefit of the various companies where I have worked. It is my hope that everyone reading these pages can take some insight from this account of one engineer's career, whether an experienced engineer, a recently graduated engineer, or a person interested in what an engineer's life might be like. I hope you enjoy it as much as I enjoyed recalling and writing down these memories.

ACKNOWLEDGMENTS

I wish to acknowledge the omnipresence of Almighty God and His support for me throughout the writing of this book. I wish to express my gratitude and thanks to my wife for her understanding and tolerance. I also would like to express my thanks to the many people who helped make my career successful and who provided me help and understanding along the way. My success in engineering is a direct result of the people I worked with and especially those who worked for me. I could not have accomplished what I did without their support, hard work, and dedication. Hiring and maintaining a loyal and devoted crew of engineers, designers, and construction coordinators was the key to any success I enjoyed. My appreciation also goes to those who corrected and redirected me when it was needed to point me in the right direction.

CHAPTER 1

Background

I grew up in a rural part of Newton County, Mississippi, in what would be considered a poor family by today's standards. We did not know we were poor, and so it did not make a difference. I had a loving family, plenty to eat, and enough work to do to keep me out of trouble. Looking back, I consider myself more fortunate than most.

We had a small farm that would not survive in today's world, but it provided us with the necessities of life and then some. I did just about every chore on the farm as expected when growing up. My father and grandfather raised what we called truck crops. These were vegetables that were grown, gathered, and taken by pickup truck into the nearby towns for sale to the local grocery stores. It was hard work, but I was expected to do my share. We raised peas, tomatoes, watermelon, cucumbers, okra, cantaloupes, butter beans, corn, etc., for sale to the grocery stores. Most of the vegetables and fruits had to be planted very early in the year in order to meet the early market demand. This required the seeds to be planted in individual containers (particularly the watermelon and cantaloupe) and protected from the cold until they could be transplanted into the fields. Tomato seeds were planted in large washtubs with soil and covered with glass to keep them warm through the cold nights. On sunny days, the glass would give the plants the sunshine and warmth needed to grow.

Once the plants were transplanted into the fields, they had to be covered with buckets and cans at night to protect from frost, and then the buckets were removed the next morning. This procedure could last for a few days or a few weeks. I helped to cover the plants but missed out on most of the uncovering of the plants because I was in school. When the plants started to grow, they had to be cultivated and hoed to keep down the grass and weeds and to provide aeration for the soil. The tomato plants had to be checked for worms and the worms removed or they would eat the plants and the young tomatoes. The watermelons and cantaloupes had to be cultivated, but since they produce long runners, it was not practical using a tractor.

We plowed them as well as our small garden with a mule, and I learned the difference between "gee" and "haw." The other plants, such as the peas, could be plowed with a tractor and cultivator until they put out runners. Once the tomatoes and cantaloupes were ready to gather, we went to the field with the tractor and trailer and gathered them. I learned to tell if a watermelon was ripe by thumping and looking at the curly part of the stem where it was attached to the melon.

Then came the tomatoes, which had to be gathered every other day. I laugh every time I hear "vine-ripened tomatoes" because we never let them fully ripen on the vine. If we did, they would be soft and subject to damage while handling. Once tomatoes start turning ripe on the end opposite the stem, they will fully ripen in a few days. This time is needed to transport to the stores and give them a few days to sell before turning too ripe. Today most tomatoes available in the supermarkets are grown on large farms and are pollinated to keep their ripe appearance and last for days or weeks while being transported and kept on store shelves.

After gathering, we had to wash or wipe the tomatoes clean and cull any that were not acceptable to sell. The tomatoes were spread out on newspaper overnight to cool and then put into baskets to carry to town the next day. The watermelons and cantaloupes were treated likewise. The peas had to be picked every other day on

ENGINEERING

Monday morning, Tuesday afternoon, and Thursday afternoon for carrying to the stores on Monday, Wednesday, and Friday. We did not pick on Sunday afternoon but rather got up and into the field when daylight came. In the early summer, this was usually around five a.m.

After picking the peas, they were spread out on my grandfather's wood porch and water was poured over them to cool them before putting into baskets to carry to town. There were times that the stores could not use the produce and did not buy. That meant that whatever did not sell was either canned or put up in the freezer for our own use. This kept me busy a lot of afternoons in the summer either shelling peas (we did not have a fancy pea sheller but relied on our fingers and thumbs) or peeling tomatoes.

We raised a wide variety of animals on the farm, mostly to eat but some to sell. I remember raising chickens, pigs, turkeys, peasants, quail, and beef cattle. I also learned to milk cows at an early age and got up before breakfast to feed cows and milk a few just for our own use. It took me a long time to get accustomed to milk that was homogenized and pasteurized when I started school. My father and grandfather had a small dairy when I was very young and converted over to a beef herd as I was growing up, but we kept a few milk cows for our use.

Even the beef cows required a lot of work. They had to be fed ground-up feed as well as hay. They required salt blocks to be put out so they could get the right nutrients. Repairs to the barn, stables, fences, and gates were required regularly. In times of extreme cold weather, we often had to go to the creek where the cows got their water and bust up the ice so they could drink. Anytime one of the cows was about to have a baby calf, we had to keep close watch to make sure the mother cow did not hide her calf somewhere in the woods. Some of the time the cows had trouble having the calves and needed help. I did more things to and for animals on the farm than I care to remember, but today someone would probably call a veterinarian for most of these things.

In support of all these things, we had to raise hay for the cows; corn for the cows, pigs, and chickens; and food for us to eat such as peas, corn, okra, cucumbers, potatoes, sweet potatoes, onions, cabbage, butter beans, string beans, tomatoes, watermelons, cantaloupes, squash, turnips, turnip greens, mustard greens, garlic, and strawberries. We took the dried corn, shelled it, and took it to a grist mill to get it ground up for making cornbread. If we wanted chicken, we went out in the yard and caught it. The chickens also provided a continuous supply of eggs. If we wanted pork, we would go to the smokehouse and get ham, bacon, etc.

When I was very young, we did not have a freezer, so everything had to be canned or smoked; but as I was growing up, we could put the pork or beef in our freezer to use whenever. My parents would buy items from a traveling merchant truck that came by about once a week. It provided things we could not produce on our own such as flour, coffee, tea, sugar, kerosene oil, etc. If you did not have the money for your purchase, you could trade the merchant eggs for it, and he would sell the eggs down the road to someone else. Since we raised most of what we ate, we did not need a lot of money to survive.

My parents married in 1939 at a very early age. My father married my mother (who was only fourteen years old) on his eighteenth birthday. I had an older brother born in April of 1942, and I came along later in December of 1948. In between there was World War II where my father served in the Army in the European Theater, was captured, and spent four months in a German POW camp. He had medical issues for the rest of his life due to his time in the service. After coming back home, he took some agricultural vocational training and began farming the land and helping my grandfather with the cows and small dairy. He purchased a new 1952 Ford N tractor with equipment to farm the land.

I recently sold the original tractor, and it still ran good even though it was ready for retirement. I had learned to drive the tractor and use it to bush hog the pasture, disk the land, run a section harrow over the newly plowed ground to smooth it out, or pull a trailer loaded with vegetable or firewood. I helped my father change the settings on the cultivator from planting to cultivating to laying by the corn and the width of the tractor wheels from wide to use as a double-row planter/cultivator to narrow for pulling the trailer or for bush hogging. Soon this was my responsibility, and I did it all myself.

The house I first grew up in was a small wooden house with no indoor plumbing and no water. We had what you would call an outhouse, and we toted buckets of water up the hill from my grandfather's house (about 150 yards) from a bucket-drawn well. When I was about five years old, my father built a new 2BR, 1B house in front of the old house and had a well drilled and we had indoor plumbing. Previous baths were in a number 2 washtub in the backyard, so when we began to take showers in the new bathroom, we were very conscious of water usage. I remember turning on the shower and getting wet then turning it off to lather up with soap and then turning the water on again to rinse off. The house had two bedrooms: one for my parents, and one for me and my older brother. There was a living room complete with a tin wood-burning stove for heat and a kitchen/dining area combination, a bathroom, and a small back porch. It was small, but it was everything we needed.

I started to school in the first grade when I was five years old (there was no kindergarten), and the school bus picked me up in front of my house and dropped me off at my house in the afternoon. There was always work to do when I got home or on the weekends. My father planted grass and established a lawn around our new house, and as soon as I was big enough, it was my job to cut grass. My older brother who was six to seven years older than me had cut the grass, but I soon inherited the task during the summer. There was always work to be done on the barn and fences as well as taking care of the crops we grew during the summer. During the fall and winter, we went down to the creek and cut down hardwood trees and split them for use as stovewood and firewood. The cows had to be fed extra during the winter because there was no grass for them to eat.

This is just a brief introduction of where I came from and how some of my values were set early in life. I consider myself fortunate and am thankful to God for all the blessings I enjoyed while growing up. In the next chapter, I will attempt to show how this introduction to life began my early training as an engineer even though I did not realize it at the time.

CHAPTER 2

Training

My training as an engineer began at a very early age even though I did not recognize the situations I dealt with as being engineering concepts. Wikipedia defines engineering as "the application of mathematics, empirical evidence, and scientific, economic, social, and practical knowledge in order to invent, innovate, design, build, maintain, research, and improve structures, machines, tools, systems, components, materials, and processes."

I will have to admit, until formally trained as an engineer, I had to rely on the practical knowledge rather than the application of mathematical or scientific principles. My older brother was practically gone from the house by the time I became a teenager, so a lot of my innovations and experiments were conducted on my own.

There was a ditch between my house and my grandparents' house that was dug around the hill to divert some of the water during heavy rains. It was normally dry and was about eighteen inches deep and about three- to four-feet wide. Whenever it rained, I dammed up the ditch and created a small shallow lake in which I could float boats and play. The boats were typically carved out of large chunks of pine bark from a nearby tree. From this experience, I learned that water will penetrate a levee that has a high percentage of sand and needs clay to be stable.

I also experimented with ditches containing saturated sand after a rain and discovered that when the sand was disturbed, it would go into a temporary quicksand condition and would lose its ability to support whatever was on top of it. This I would learn later on in soils classes and also the effect of earthquakes on certain types of soils. I was fascinated by the flow of water, the levee that held the water back, the composition of the levee material, the effects of rapidly flowing water on erosion, and the quicksand condition in saturated soils. I would learn later in more technical terms what was happening and why, but I was already gaining valuable field experience.

Usually around Christmas time or around the Fourth of July, I had access to fireworks, and I took advantage of them. There was an old roadway ditch bank that was perfect for placing firecrackers in the bank and lighting them to see how much soil could be removed. I found out later on that this is a normal practice for removing rock material on a much larger scale.

I learned early on why cultivated land on hillsides must have terraces to divert the rainwater around the hill rather than letting the water flow down the hill and cause extensive erosion. I guess this was early training in surveying. I learned about corrosion early on also. You do not place fertilizer in metal buckets and leave it there because it will corrode. To keep the terminals on a tractor or car battery from corroding, coat the terminals with lubricating grease. If the terminals do have a small amount of corrosion, it can be removed by pouring Coca-Cola (or any other carbonated drink) over the terminals. I would later learn in chemistry class what happens when you put a base and an acid together.

I learned a lot about the simple machines and tools that are a bases for a lot of engineering. The simple machines are levers, inclined planes, wedges, pulleys, wheel-and-axles, and screws. I used the lever to jack up the tractor. Using a combination of frames and chains, the special tractor jack could be placed under the front and rear axles of the tractor and connected to the hydraulic lift. When the lift was raised, the chain tightened, and with the advantage of the lever, it lifted the entire tractor off the ground so both front and rear tires could be repositioned.

The inclined plane was used to load and unload a fifty-five-gallon drum onto a pickup truck to go to town and get gasoline for the farm tractor. It made it easier to load the empty drum, but it was essential for off-loading the full drum, which would weight over three-hundred lbs. I used the wedge every time I swung the axe while splitting wood and also saw the benefits of using a wedge to adjust and line up things for connecting. The pulley was used to draw water from the well every time I drew a bucket of water.

When I was very young, I also watched the well being dug using large sections of precast concrete pipe and a bucket-and-pulley system for removing the dirt from the well. I observed the wheel and axle in action while watching the planters mounted on the cultivators. There was a chain that ran around several sprockets connecting the fertilize dispenser, the seed dispenser, and the steel wheel that rolled on the ground causing the chain to move. The amount of fertilizer and the spacing of the seeds could be modified by using different-sized sprockets and idle sprockets. I also used the wheel-and-axle concept while using a corn sheller that consisted of a steel plate with tiny fingers to grab the corn ear and rotate it while removing the kernels of dried corn. The plate was connected to an axle and handle for rotating. The screw was used every time I adjusted the cultivators for a different spacing and size of plow. The screw was used to jack up a vehicle, bolt/unbolt wheels, change the spark plugs on the tractor, truck, or car or anything else that needed to be bolted down.

I learned other principles of engineering while on the farm. I helped my father to plant crops such as corn using the planters on the farm tractor. It was my job to keep the seed and fertilizer hopper filled up while he was operating the tractor. It was not difficult to keep the seed hopper filled, but the fertilizer was harder to do. We would usually carry the fertilizer to the edge of the field with the truck or to the ends of the rows with the tractor, but on long rows, the fertilizer had to be replenished whenever needed. I learned very quick that if a full hopper could go two-and-a-half rounds, then it was to my advantage to fill it up after two rounds rather than have

ENGINEERING

to carry the fertilizer to the far end of the rows to fill up before it ran out. Those fertilizer sacks weighted one hundred lbs. And even though I could handle them, I preferred not to. I used a little industrial engineering that I called common sense through observation and application.

I practiced good engineering concepts when constructing things on the farm. I helped build a pole barn using good bracing techniques, used wedges for inside door locks, installed rain gutters over doors to prevent water from entering the door openings, used good drainage practices, and learned a lot about span distances, the proper way to turn sawn lumber for the most strength, and how to brace doors to prevent sagging. I learned how to construct a barbed-wire fence using braced sections of fence posts and wire pullers to stretch the wire. I learned how to make flexible wire gates with chain latches that could be easily opened by someone with two hands but not easy for cattle to open. I helped construct and repair simple timber bridges across drainage ditches in order to access the fields with the tractor.

Work on the farm was hard, but it had rewards. Whenever the work was caught up, we could go fishing in either the creek or the pond, and both had fish good for eating. We would also get to go swimming either in our pond or a neighbor's pond, usually about once a week in the summer. I had a bicycle that I logged many miles on and could explore the surrounding area. We could watch TV after our chores were done. I remember our first TV, which could get three channels if we were lucky. You walked up to the TV, turned it on, set the volume, chose which of the three channels you wanted to watch, and then you set down. When you were through, you walked back to the TV and turned it off. There was no remote, no educational TV, and no cable; but it was all we needed. I had a collie dog to keep me company and the area around me as far as I could walk or ride by bicycle to explore and learn from nature.

My love of the soil, water, and structures, etc. on the farm naturally drew me to engineering. In school I had taken courses like many of my classmates in agriculture, was a member and officer in the school FFA program, and obtained the status of a state farmer.

I placed second in a public-speaking contest in regional competition (the first-place winner went on to capture first place in the state competition). My topic was "Water," and I demonstrated the importance, the preservation, and the need for good water management for future generations. This speech was appropriate for an agricultural audience as well as a civil engineering audience.

Even though my father and grandfather worked hard and earned their livelihood on the farm, I began to lean more to engineering. I had a very strong background in mathematics and enjoyed science, especially physics and chemistry. Based on my love of soils, water, dams, structures, and buildings, I began looking at different career fields in engineering. By the time I finished high school, I had decided to attend college and work toward a degree in civil engineering. I did, however, have a difficult time explaining to some of the older members of the local church that becoming an engineer did not mean that I would be driving the train.

CHAPTER 3

Formal Education

    I was fortunate that literally across the street from the high school was a two-year college, East Central Junior College, that offered a very good pre-engineering curriculum. Since money was tight, I could enroll in the junior college and stay at home and reduce the cost of my first two years of college. While in my senior year of high school, I secured a part-time job driving a school bus for the high school and continued while attending East Central. This provided me with transportation back and forth and enough spending money while I was there.

    College was certainly different from high school. But since I was a day student, I did not get totally immersed in the college life. I traveled to classes every day and stayed home each night. While staying at home, I continued to have chores and responsibilities and continued to develop my practical engineering skills. My classes were usually between eight a.m. and three p.m. And since I could not leave until I drove the school bus home at 3:15 p.m., I used the times between classes to work on homework and studying for classes and tests. The requirements for college classes were much more than in high school, and even though we had less time in class, we had a lot more outside work to do.

After I got home at around 4:15, I started on my chores: feeding cows and pigs, milking cows, gathering firewood, putting out hay, and whatever else needed to be done. Then after eating supper, I would begin again on homework until bedtime. The next morning, I got up, ate breakfast, fed and milked cows, and got ready to drive the school bus into town and start another day at college. I got a little break on the weekends after my homework was done, but for the most part, this was my schedule for the college semester.

The same month I started college, I also enlisted in the local Army National Guard. This break from college would give me the opportunity to develop my military education and discipline. Even though I was only seventeen (my parents had to sign for me to enlist), I wanted to be able to serve my country without having to have a long break in my college classes. By enlisting in the National Guard, I would only have to be gone for about four months for basic training and a military specialist school and then serve a minimum of six years in the Guard. The local Guard had a lot of my friends from school and other people I knew including my older brother. I was ordered to basic training in February, so I did not enroll in the spring semester at East Central. I left around midnight from a local bus stop on a Trailways bus headed for Fort Polk, Louisiana. I did not know what to expect and was greeted by a completely different lifestyle for the next four months.

Since most of my teen years were spent at home with no siblings close to my age, it was strange to be thrown into a group of about fifty people that I would spend the next several months doing everything. Military life was not easy but it was very structured and disciplined, and if you did everything requested of you the way it was expected, you could survive. I often heard the statement "There is a right way, a wrong way, and the Army way." There was a scheduled time to go to bed, time to wake up, time to eat, time to exercise, time to train, and a time to do everything.

Lights out was at nine p.m., and lights were turned back on at 4:30 a.m. to get up. The Army guaranteed us seven-and-a-half hours of sleep. We were also required to get up on a rotating shift during the night and look for fires in the building for a thirty-min-

ENGINEERING

ute period. We were awakened a half hour prior to our shift so we could get fully dressed with pressed fatigues and spit-shined boots. The barracks we were in had been hastily built during World War II and had been documented to burn completely down in about twelve to fifteen minutes, hence the requirement for the fireguards.

This duty came around about every three to four days, so a typical night would go like this: lights out at 9:00, fall asleep at 9:30, wake up at 11:30, fireguard from 12:00 to 12:30, go back to sleep at 1:00, and wake up at 4:30. I think this was more harassment than guarding for fires since no one was watching the barracks when we were gone during the day. We considered ourselves lucky because we were only a couple blocks away from Tigerland, where the troops were training to go directly to Vietnam. They were kept up until one or two a.m. in classes, and most of the time, they woke us up at four a.m. running by our barracks screaming.

Exercise was a way of life, and we were constantly given the opportunity to improve our health. We usually had about an hour of exercise in the mornings to start off that included side-straddle hops, four-count push-ups (two push-ups—up-down, up-down), eight-count push-ups (two regular push-ups starting at the position of attention, then squatting, one push-up, then back to attention, and repeat for the second push-up), and a multitude of other exhausting exercises. Our drill instructors were not very good at math. We were only required to do twenty-five push-ups at a time. As an example, we would do fourteen push-ups, and then the drill instructor would ask, "How many push-ups is that trainee?" and we would reply "Fourteen, drill sergeant."

He would then say, "No, I think it is six," and we would start again back at six. This would go on for two or three times, and we would usually do at least fifty before stopping with our maximum of "twenty-five" push-ups. Sometimes the drill instructor would ask if we had had enough, and of course, if we replied yes, he would disagree and give us more. We finally learned that to stop the exercise, we should beg for more, which we did. And usually after granting our wish a time or two, he would stop and move on to something else. The most push-ups I remember having to do in a row was 175, and I believe the drill instructor stopped only because he could see I was able to continue. Looking back on this, I realized that this was excellent training for my later career in engineering to be able to handle those who made promises and did not keep them such as clients, salesmen, vendors, and sometimes, management.

We also were allowed to exercise before eating each meal, while moving to and from the classrooms, and anytime we went to the field. I remember one time running in full combat gear, combat boots, and with a rifle over my head for several miles. I guess we were in a hurry to get there and did not want to get our rifles submerged in case of an unexpected flash flood. We marched or ran everywhere we went while singing songs or calling cadence. We trained on hand-to-hand combat skills; the use of a bayonet; how to throw a grenade; nuclear, biological, and chemical warfare; and the required protection. We also spent a lot of time on our assigned weapon learning how to shoot proficiently and clean and disassemble it into over a hundred separate parts and then reassemble it in timed exercise.

On the weekends, we had a different schedule. On Saturday morning, we took a PT (physical training) test that consisted of a run-dodge-and-jump, a grenade toss, a forty-yard low crawl, parallel bars, and a one-mile run. Each exercise required something different. The run-dodge-and-jump required quickness and agility. The grenade toss required proper form and accuracy. The forty-yard low crawl required guts and determination. Canvas mats were spread on hard-packed clay, and each soldier had to crawl on his belly without lifting his chest off the ground for twenty yards down and twenty yards back. It was grueling and often rubbed holes in the fatigues and wore the boots down to the brown leather. It had to be done in under twenty-five seconds to qualify.

The parallel bars were essentially ladders laid out horizontally, and you had to swing from bar to bar and turn around at the ends without slipping off. The one-mile run was done in full uniform with combat boots. None of this sissy stuff like running shorts or athletic shoes allowed, and it was usually done after all the other exercises. We got to do this every Saturday, and we did improve over the course of the eight weeks in basic training. I learned that not everything you

ENGINEERING

attempt to do is totally successful on the first try, but if you apply yourself, you can accomplish whatever you want to do. And this is especially true in engineering.

On Sundays we sometimes had duties and details but also had some free time to polish our boots and brass, write letters home, etc. We were still required to get up and dress in the proper uniform for meals but did not have classes. We were allowed a one-hour period on Sunday morning to attend the church of our choice. They had a Protestant service, a Catholic service, and a Jewish service. Whoever was going to each service had to be marched to and from the service, but once inside, it was the one hour that you did not have to look over your shoulder to see if the drill instructor was watching you.

Our training was not all physical; there was plenty of mental applications. We studied about code of conduct, learned to discipline ourselves, learned to follow instructions, learned respect for our supervisors, and learned to work as a team. At the time, we probably did not think highly of our drill instructors, but I did understand what they were trying to do and respected them for it. It was probably one of the most important things that helped me get through basic training. Later I learned to fully appreciate what they had done for me to help me develop discipline and skills to be used as an engineer. Throughout my career, I always looked at military training and experience as a very positive attribute when evaluating potential applicants for a job.

After my basic training was completed, I was assigned to a supply school at Fort Polk to receive training for my military career. This assignment was much more relaxed and businesslike. I still had to meet strict Army rules and regulations, but most of my days were spent in classes learning to be a supply specialist. We were still told that since we had completed our initial training, we could be shipped directly to a war zone if needed since we were currently on active duty. I had learned that on Saturday morning, if you could manage to escape from your assigned barracks, they could not get you for a detail. I had gone one Saturday to the post swimming pool, and once I was in my swim trunks, I was not easily identifiable. While at the pool, I heard the announcement on the radio that Israel had attacked

Egypt, and I was sure that I was about to be sent to the Middle East. Fortunately for me, the conflict was soon over. This was the Israeli Six-Day War and was over before I was off active duty.

I was released from active duty after my supply class was over and was once again serving my time in the National Guard on weekends and the annual two-week training period in the summer. While in the Guard, I was activated several times—a couple of times for riot control, which fortunately, did not turn into riots. I missed a couple of activations because I was too far away at school to be able to get to the unit before it pulled out. I did, however, get to stay on the Mississippi Gulf Coast, all expenses paid, for a week in August of 1969 due to Hurricane Camille. It was the worst thing to hit the area and left lots of destruction in its path. Our days consisted of manning roadblocks to keep sightseers out of the area, preventing looters from entering damaged and abandoned buildings, adding Clorox to disinfect emergency drinking water supplies, delivering water and food supplies, and manning details to look for dead bodies. Our detail found in a pile of debris a terrible smell that could have been a dead body. We were happy to discover that it was only a dead deer.

The thing I remember most was the damage the wind and water had done, especially the water. The wave action had lifted huge sections of concrete roadway weighing many tons on the main US Highway 90 bridge over the Biloxi Bay and stacked them like dominoes. I also saw a two-story wooden framed house that appeared untouched on the top story, but the bottom story was completely gone except for the wooden columns holding up the top floor. This was a field trip I would never forget, and I have thought of it often over the years when designing for wind- and wave-action loads on structures.

I returned to East Central after my active duty in the Army in the fall, and because of the sequence of classes being taught, I attended for another two years to gain my associate of arts degree in pre-engineering. I was able to take a lot of electives during this period but concentrated on the core engineering courses I would need to transfer to a senior college. The first semester before the Army was mostly general courses required of all students such as history, English literature, mathematics, and science. I always enjoyed math classes, so I eagerly took classes in trigonometry, differential calculus, integral calculus I and II, and differential equations that formed the basis of the equations and calculations used to solve engineering problems. I took science classes in chemistry and physics, which was the basis of understanding the physical and chemical properties required for the practice of engineering. I gained insight into using a library for research and verification of facts and not just a place to check out books to read or magazines to browse.

I began to take courses directly related to engineering. A graphics class was used to develop 3D visualization of objects and to look at all objects in plan, front, and side views. Today 3D objects are easy to visualize by simply rotating a 3D object to look at all six different views. I took mechanical drafting in order to learn how to letter, draw, and produce drawings that would accurately portray to the fabricator or installer exactly what was to be built or fabricated based on an engineered design. This was the real core of my engineering. Today we have progressed through manual lettering of text and notes to using mechanical aids such as Leroy lettering. Now we simply type into the computer, and the CADD (computer-aided design and drafting) inserts the text onto our drawings and documents.

I was introduced to electric-circuit theory from another university by use of a closed-circuit TV system. Today this would be taught on an online course; but when I attended college there was no internet, no handheld phones, no personnel computers, and for all practical purposes, no handheld calculators. I'm sure there may have been some around, but they were not available for the average engineering student. We had a secret weapon though: the slide rule. With proper training and practice you could multiply, divide, solve quadratic equations, solve for square roots, etc. accurately to three significant figures. It was easy to spot the typical "engineering nerd" because they probably had a slide rule strapped on their belt loop and a pocket protector in their shirt pocket. Add a pair of glasses, and the uniform was complete.

It was not that the engineer was trying to be nerdy, but by necessity, most engineers are logical. The belt loop for the slide rule made it accessible and also made it easier to carry all the required books. The pocket protectors kept the shirts from getting inked by mistake. I must confess here that I did not wear a pocket protector but should have. I have ruined many shirts with an ink pen in the pocket unprotected. I am still plagued with that potential problem (old habits die hard).

There were some courses I took while at the junior college, even though not required, that broadened my education. I took additional history, speech, and speed-reading classes. The speech class helped me in my engineering career to be able to get my ideas across to the chosen audience, and the speed-reading class helped me to cover material much faster when researching and studying background material. Unfortunately, I did not learn how to speed-read complicated formulas and detailed technical discussions, but it did help with other things. I graduated from East Central Junior College with an associate of arts degree in pre-engineering and transferred those credits to Mississippi State University to complete my bachelor's degree.

My transfer to Mississippi State was a change in many ways. I became a part of a university that was more than ten times larger in the number of students, buildings, classrooms, instructors, and traffic jams. I was assigned to a dormitory and began my life as an on-campus engineering student. In spite of my number of transfer credits from East Central, I was considered a junior-level student. My roommate was an engineering student but at the sophomore level, so I did not have any classes with him. In fact, I don't think he attended very many classes at all. When I left for breakfast in the mornings, I would sometimes meet him coming in from the night before. When I returned to my dormitory in the afternoon, he was usually playing racquetball or some other activity. Some days I did not see him at all. After one semester, he flunked out, and he was gone. I don't think he had the desire or drive to be an engineer.

The second semester I had the room all to myself since no one else was assigned as a roommate. My transfer credits were good for most all my freshman and sophomore classes and a few junior-level classes, but I had to take a few classes that everyone else took in their sophomore year. I was required to take a one-hour course in slide rule even though I had been self-taught and had been using one for about two years. I learned how to do even more with proper training.

I settled in quickly and began to take the core engineering classes such as engineering mechanics, strength of materials, soil mechanics, water supply and wastewater control, surveying, highway engineering, hydraulics, structural steel, reinforced concrete, structural analysis, numerical analysis, engineering economy, photogrammetry, highway materials, and numerous lab classes coinciding with soil mechanics, surveying, hydraulics, and highway materials. There were very few electives since I had already taken most of them in junior college. The later part of my senior year, we worked on team projects that incorporated the classes we had completed on soils, concrete, structural steel, etc. There was a choice on which major area of civil engineering to concentrate. You could concentrate on environmental, highway, structural, soils, etc., and I chose the structural major.

In order to pay for part of my college expenses, I worked part-time jobs the entire time I was in college. I respected anyone who could have outside activities, a part-time job, take a full engineering load of classes, and get respectable grades. My first job was through the university, and I worked part-time for the maintenance crew at a USDA research lab. It paid $1.50 per hour, and I was glad to get it. I worked changing light bulbs, cleaning and replacing filters, trimming shrubs, and helping to repair equipment used in the lab. I learned to replace electrical outlets, test and maintain the chemicals in the facility cooling tower, operate a sheet metal break, spot weld, and solder sheet metal. I had a great supply of tools to use, and I made the most of them while there.

After the first semester the head of maintenance offered me a job under a different program, and my pay increased to $2.00 per hour. I kept that job for the time I was in college. Even though most of my work was during normal daytime hours, my supervisor allowed me to have a key for access to the building, and I knew where the key cabinet was, thereby giving me access to the entire facility. I had certain tasks that I was expected to perform, and sometimes, they were handled during off-hours.

I understood later on the significance of the chemicals to treat the cooling tower water in my water treatment class and had a very practical introduction to electrical circuits, mechanical equipment, and airflow in HVAC systems as a good introduction to engineering problems and solutions not always found at home on the farm. I continued to learn and absorb information and knowledge I could use later on in my engineering career.

My senior-level courses were much harder and demanded more time for study, research, and understanding. Each professor acted like their class was the only one you had, so they piled on the work. My social activities consisted of watching reruns of the *Wild, Wild West* TV show in the dorm lounge with a group of other students and occasionally going out to eat. I usually came home most weekends, so there was not much extra time to do other things. I did stay on campus for football games, basketball games, baseball games, and an occasional concert; but most of my free time was spent working on homework, writing reports, or studying for exams. Most classwork was accomplished on a priority system. If you had two tests due in two days and a report due the next day, you worked on the report and studied for the tests the following day/night.

Much of my reading and studying for exams were done at night. Sometimes I would go to the library to study, but I found it much too quiet. And every time someone softly walked by, it got my attention and broke my concentration. I studied sometimes in the food area and on couches and tables where there was more noise, and I found that I could tune it out and concentrate on what I was studying. Sometimes at night, when I was too drowsy to stay awake, I would walk around the campus and stop on benches under streetlights and keep moving from place to place to keep me awake. Sometime this was at one or two in the morning, but I had lots to do keeping up a full schedule.

I always worried before the exam but never during the exam. I gave it my best, and that was all that I could do. My grades could have been better if there had been fewer classes in a given semester, there was no part-time work, and there had been no activities such as ballgames or concerts. It did, however, give me the ability to work on multiple tasks simultaneously, prioritize my tasks, and develop time-management skills. All these would prove to be vitally important in my upcoming engineering career.

The classes I enjoyed most were the structural classes such as structural steel, reinforced concrete, structural analysis, and numerical analysis. I also enjoyed the classes that supported structural foundations, such as soil mechanics and highway materials, and general classes, such as surveying and photogrammetry. Photogrammetry is closely related to aerial surveying. Photos are taken of a given area from different angles using an airplane flying overhead. The developed pictures are observed using a viewer that lets each eye concentrate on the same site from a different angle and, when in proper focus, allows the observer to view the photographed scene in 3D. This is similar to the red-and-blue glasses used at movie theaters to view 3D movies. I was able to focus each eye separately on the two pictures that I could see the 3D images without the viewer. This is similar to the color scenes sometimes placed in Sunday newspapers that if you focus your eyes properly, you can see the 3D images. This technology is the basis for obtaining aerial surveys that can produce large topographical maps of areas at economical costs, and I have used maps produced by this method many times in my career. There were, of course, some classes that were not my favorites; but I came to realize in later years that they were just as important to my engineering education and career.

The labs that I took were equally challenging but provided me with insight to hands-on experiments and testing that reinforced my lectures in the various subjects. One of them, a highway materials lab, gave me real-world experience on concrete and asphalt mix design as well as testing concrete cylinder breaks. This helped me to understand the concrete break results I would receive on the engineering projects during my career and why they were so important.

My soils lab introduced me to many of the testing procedures that would be required in specifications that would be issued for future projects. In the hydraulics lab, I learned the importance of fluid flow in pipes and open channels and the effects of friction on the fluid flow. I studied the differences between laminar and turbulent flow and gained an appreciation of pressure and gravity flow systems. I developed surveying skills in the field in order to determine the location and elevation of existing facilities and the proper way to locate and specify the location and elevation of new facilities.

I do not want to give the impression that college was easy or that I just breezed through my classes. They were very hard and demanding, and I often did not score 100 percent on my exams or homework assignments. That is all a part of learning, and I think it helped me to understand that I should always strive to do better and never give up on assignments or life in general. I learned that anything worth having or doing was worth working for, and sometimes, you have to work on multiple tasks simultaneously to accomplish your goals. Later on in my civil engineering career, I would tell those working for me that college gives us the skills to analyze problems when presented with them but real life requires that we often have to determine what the problem is before we can begin to solve it, and that takes experience.

CHAPTER 4

A Summer Job

I graduated from Mississippi State University in May 1971, earning a bachelor of science degree in civil engineering with a major in structural engineering. I accepted a summer job as an engineering intern for a major oil company working in the offshore construction division with their office located in New Orleans, Louisiana, and with projects located in the Gulf of Mexico. This was also my first experience living away from home other than my time in Army basic training. I was twenty-two years old living in the big city of New Orleans in my own apartment and was responsible for everything that happened to me, both the good and the bad.

After accepting the summer job, my first task was to find an apartment that was relatively near work that I could afford. I located a one-bedroom apartment in Metairie, Louisiana, which was adjacent to Orleans Parish and the official city of New Orleans. It was only about a thirty-minute drive to work, so I moved all my possessions in my car and set up temporary residency. I remember the utilities being included in the apartment rent, but I had to pay for a telephone separate. Because I had no history of credit except things I paid for while at school, the telephone company required a large deposit before I could obtain long-distance services. This was prior to cell phones, and if you wanted to talk with anyone on the phone, you had to have a landline or go down to the corner and use a pay phone. Since I was away from home and knew no one except the people I worked with, I made several telephone calls during the month. As soon as I reached my deposit limit, I was required to go to the phone company and pay my bill in the middle of the month to continue my service.

My first day at work was a traffic nightmare, and then I began to realize that every day was a traffic nightmare. I was lucky enough to have the use of a company car. I was told to take the bus to work the first morning and then I would drive the company car home the first day. My traffic problems were not bad catching the bus. New Orleans has a decent bus service, and many people get around using either a bus or a trolley. After receiving my company car and laying out my route on a map, I left work and immediately was in the middle of heavy congested traffic. Every time I approached the street, I needed to turn right, but the traffic was backed up and I could not turn. When the light changed to green, the traffic from the other direction had the intersection blocked, and I could not turn. After going through about three light cycles with no success, I blindly struck out in a new direction that would take me in the general direction of my apartment. This was no better because in New Orleans some of the streets are discontinued for a couple of blocks, and even being on the correct street going the correct way may not get you there. After about thirty minutes, I turned back to downtown where I worked on Tulane Avenue, near the French Quarter, and found a new route to get to the apartment. It took me about one-and-a-half hours the first day to get back to my apartment.

That first drive home almost turned into a major fender bender when another car cut across my lane out of a blind spot and almost hit me. Thank goodness for good brakes. If you had a car in New Orleans, you were required to have it inspected for brakes. Your vehicle was actually put on a machine and tested for the brake capacity before you could get an inspection sticker. There was not so much concern over mirrors, glass, etc. but the brakes were inspected rigorously for a good reason.

ENGINEERING

My company car was a full-sized Ford sedan with an automatic transmission and a heater/air conditioner and nothing else. The air conditioner was a must for the New Orleans temperature and humidity. If you wanted music or news, you bought a portable radio and placed it on the seat. The main purpose of the radio was to check for traffic and wrecks in the city. It was a daily occurrence to be delayed by a wreck unless you could avoid the area by taking a different route. It did not take long to learn the best ways to go and the best alternate ways to go to get around the city.

My first day at work consisted of meeting my bosses and the other engineers and designers in the group and learn what projects they were handling. I was in the offshore construction division and was hired along with an engineer from the University of Missouri at Rolla to follow offshore soils exploration in the Gulf of Mexico. When we were not on assignment in the Gulf, we helped on other minor projects and did anything else requested of young engineers that were eager but had no real experience. We were often sent out on assignments to the fabrication yards, dock facilities, and offshore platforms to be representative for the company, relay information back to the main office, and be the eyes on the ground for the projects we were assigned. The reason we needed the company car was so we could visit at any hour of the day and any day of the week as needed. I remember leaving at 4:30 in the morning to visit a fabrication yard or driving to Venice, Louisiana, to meet a workboat headed into the Gulf at 12:30 a.m.

The trip to the boat dock in Venice required me to leave my apartment around ten p.m. at night in order to meet a boat leaving the dock at 12:30 a.m. It was an uneventful trip there, and I arrived in Venice and located the dock area with no problem and had about twenty to thirty minutes to wait. Since there was no one there when I arrived, I assumed that anyone meeting the boat would be arriving just in time to board the boat and leave. When it was about ten minutes until departure and no one was there, I began to have doubts and discovered that there was another dock and I was at the wrong location. I left to locate the correct dock, but by the time I arrived, the boat had left. There was nothing else to do but to head back to the apartment.

On my way back to New Orleans, I passed through the town of Port Sulphur at around two a.m. There was a posted speed limit of 30 MPH, and since it was late and there was no one else on the road, I slowed down to about 40 or 45 MPH and apparently woke up the local police who were waiting for me. I could not complain since I was legally speeding, so I gave the officer my license and registration as he requested. The registration was in order and so was my Mississippi driver's license, but I had not obtained a Louisiana driver's license for the summer job and the company car had a Louisiana vehicle tag.

The officer stated that since I did not have a Louisiana driver's license and was driving a car registered in Louisiana, he could not give me a ticket and his only option was to let me go or throw me in jail. I realized later that he was just pulling my leg and needed something to liven up his boring night shift. If he had wanted to, he could have easily given me a ticket. I was getting ready to spend the rest of the night in jail when he told me to leave and drive carefully and stay under the speed limit. To say I was a bit naive was an understatement, but I did arrive back at my apartment safe and sound and got about three hours of sleep before heading back to work.

Because I had missed the boat, another mode of transportation was required to get to the drilling platform in the Gulf. I was scheduled to ride on a crew-sized helicopter to the platform that morning with some other people. It was my first helicopter ride, and it only took about thirty minutes to get there versus the two-hour car ride and the two-hour boat ride I was scheduled for originally. I guess I was lucky because that particular helicopter ride had crashed the day before on the same route.

Most of my assignments in the Gulf were either on workboats that were taking soil samples on the Gulf floor at sixty to one hundred below the surface of the water or on derrick barges used to construct the offshore platforms and move in production platforms in order to produce the oil. The reason for my employment was to follow the soil drilling rigs. In order to drill on the bottom and obtain

ENGINEERING

samples, the drilling rigs had to be cased the entire depth from the boat to the soil floor. The drill stem, drill head, and sampling bits were passed from the boat down through the casing and into the soil. This was accomplished by driving a regular truck-mounted drilling rig onto the deck of a workboat and welding down pad eyes to the metal boat deck and securing the drilling truck to the pad eyes with chains and binders.

These workboats were 155 feet long and had a cased hole in the middle of the boat from the boat deck down to the bottom of the boat hull so the cased drilling could be lowered through this opening and extended to the soil bottom. To secure the workboat, four anchors with winches were utilized to lock the boat into position. Once the boat was in position and locked down, the casing was installed, the drilling stem and drill bits were lowered into the soil, and then the sampling began. Once the drilling operation began, it did not stop until it was completed. The drilling crew was from a local geotechnical firm and consisted of two five-man teams that alternated with twelve hours on and twelve hours off. There was one engineer with the geotechnical firm to supervise and record the samples. The five-man crew slept in pup tents on the boat deck while not working. The boat had a captain and a first mate to do the anchor/winch work, the cooking, and anything else the captain needed. There was a total of fourteen people on the boat including myself, who was there to represent my company.

The first time I went out, the captain handed me a piece of paper and a pencil and asked me what I wanted to eat on the trip for the next week. He could not read but was going to buy groceries before we left. We ate steaks with all the trimmings, and I asked for strawberries (I love strawberries). The captain was in charge of the boat, but as the only representative of the client on board, I felt like I was in charge of the project locally. I had finally arrived as the engineer-in-charge (make that engineer-on-site).

I made multiple trips on the workboats drilling for soil samples in the Gulf from southeast of New Orleans at several Mississippi River outlets to south of Lake Charles near the Texas line. On one particular trip to a location approximately one hundred miles south of Cameron, Louisiana, we encountered four- to five-foot seas and an anchor chain snapped in the rough seas. Without the anchor, we could not continue the drilling and had to come in for repairs. It took most of the night traveling back to Cameron for repairs. The seas were so rough that it was impossible to sleep lying on my side, so I was forced to lay on my back to get any sleep. There was nothing in the little town as I was stranded on Saturday night and all day Sunday. I found a drugstore open, and I bought comic books to read to pass the time. Then it was back to the open Gulf for more soil exploration.

The purpose of the geotechnical work was to establish the capacity of the soil in a given area for supporting an offshore platform and drilling rig. The state of Louisiana had previously let bids for leases on tracts in the Gulf, and these particular areas had been leased by my company for the purpose of crude oil production and recovery. Depending on the soil capacity and the platform anticipated, the bottom support for the platform would be designed by our group, drawings produced and issued for construction, and the construction by the successful construction bidder would be supervised in the field.

After the platforms had been designed and fabricated, the construction contractor would move into the field with a typical five-hundred-ton derrick barge to erect the platforms. The platforms consisted of two parts: the substructure that was primarily below the water level with the top usually about four feet above the waterline, and the superstructure that is above the waterline and may rise to about eighty feet above the waterline to place the main structure above the possible storm surge during a hurricane.

The substructure is usually floated to the site from the fabrication yard by plugging the steel pipe legs of the structure so that it will float in the water. When installed, the top would be a few feet above the waterline and most of the structure is below the waterline. When the substructure arrives on site, it is connected to the hoist of the five-hundred-ton derrick barge and the plugged legs are allowed to take on water, gradually lowering the substructure until it is totally supported by the derrick barge. The substructure has steel mats attached at the sea-bottom level that will allow for water to pass through and the mats to be supported by the soil. At the correct elevation, the substructure is held in place while the supporting piles are driven and welded to hold it to the correct elevation.

It was a fascinating procedure, and I got to witness it up close and personal. After the substructure is set, the superstructure is brought to the site, usually on a derrick barge, and it is set and welded to the substructure. As the company representative, I was asked to hold a survey rod on a twenty-four-inch diameter cross member of the substructure about three feet above the waterline to verify the correct elevation. I realized this was a test for the "new engineer" and tried to look nonchalant as I walked out on the pipe with no handrail, guardrail, or other means to support me. I am sure I looked like a nervous high-wire act with a balancing pole, but I did it and moved on.

In order to secure the substructure to the supporting piles, the piles are driven through the legs of the substructure using a continuous hydraulic pile driver that stopped only long enough to cut the end of the hammered support pile and weld on a new pile section before resuming the pile driving. The piles were typically driven 150 to 180 feet into the soil plus the depth of the water requiring multiple sections of pipe to be used. This would continue for several days around the clock until completed on all legs. After the support piles are driven to their final depth, the annular space between the support legs and the support piles are grouted for continuous contact.

One of the particular platforms I worked on was a new design in which the drilling was done through the middle of the supporting legs. The typical practice before that was to have a set of guide pipes along one side of the platform in which the drilling was placed. There might be as many as a dozen holes drilled off one platform, and after reaching a certain depth, they would be drilled at an angle to spread out the area to be drilled for oil. By drilling through the legs, the overall size of the platform and the wind/wave action surface area could be reduced and the overall cost of the platform could be reduced.

After the superstructure is set and welded off the initial platform, installation is complete. It is now ready for a drilling platform to be attached, which is usually performed under a separate contract. Typically, a drilling platform was installed on the superstructure that consisted of several packages transported from another offshore platform that were necessary to do the exploratory drilling in order to reach the oil deposits below the Gulf of Mexico. This included the drilling equipment, the support facilities for the workers, a heliport landing area for transporting personnel and minor equipment, storage and maintenance facilities, etc. All the packages were welded to the superstructure and stacked on top of each other to complete the drilling platform.

After the drilling platform is in place, the derrick barge is removed and drilling begins. If successful, a contract is let for the removal of the drilling platform no longer required, transportation to a new superstructure, and the installation of a production platform. The production platform is used to receive the crude oil, separate it into various components through a piping arrangement known as a "Christmas tree," and usually feed into a pipeline underwater to pump the oil/gas to an onshore facility to process the crude oil and gas.

I learned so much more by being on-site and observing what was happening and why. I felt like I was carrying two or three years of experience back to school by my exposure to the field in only about three months. Experience in the field is one of the most critical parts of a young engineer's experience and, today, is not always valued as it should be. I gained knowledge about wind and wave action on structures and why it is so very important. I was able to observe pile driving, annular grouting, automatic welding and cutting, rigging and critical lifts, geotechnical exploration, soil sampling, underconsolidated soils, contractor work practices, and contractor/client relationships and practices.

There was time for other things as well. Watching the sunrise across a clear blue sky in the morning was a beautiful sight that you cannot comprehend when your view of a sunrise is obstructed by trees, buildings, smog, or anything else that interferes with the view. I was also able to check out the fishing. I had always heard that fish collect around the drilling rigs. On one of my trips to a rig, I purchased about two hundred feet of one hundred lb. test line, a large hook, and some lead anchors. After placing the hook and lead weights along with some bait, the line was lowered over the side of the rig and tied off so it would be about ten feet above the bottom (I knew the depth of water at the rig). The next morning, when I checked the line, it was chewed off at the end and all was gone—hook, line, and sinker. So much for catching the big fish.

While working on the derrick barges, food was served around the clock every six hours: six a.m., noon, six p.m., and midnight. This was necessary for the crews working twelve-hour shifts, but for an engineer whose responsibilities were more along the lines of observe, report, and perform tasks as requested, it was way too easy to wander in and eat every six hours whether you were hungry or not because the food was excellent. I have learned over the years that cooks from Louisiana have developed excellent culinary skills and that they can produce great food out of almost anything. If I stayed on a derrick barge a couple of weeks in a row, I would easily gain about ten pounds because I loved to eat. There was one assignment on a derrick barge that the only representatives from my company were an inspector and myself. The main chef discovered who I was, and every morning, he sent one of his cooks out to find me on the barge, no matter where I was located, to provide me with a fresh pastry. I did not let the treatment affect any of my decisions, but it sure made me a more relaxed person to be treated so special since my mama was not around to offer me special treats.

The construction workers were always testing the new guy, and on the derrick barge, it was no different. We had just set a superstructure and were almost complete with the assignment when I was asked to check and verify a weld. By this time, I knew they were testing me. But I pretended like they really did need me to check the weld, so I did. The weld happened to be on the bottom side of a ten-inch beam about eighty feet in the air with nothing under it except blue water. The weld was also about ten or twelve feet from the edge of the structure where it could be accessed. I did not feel comfortable with looking at the weld, but I would not give them the satisfaction of getting the best of me. At that time, there was no strict requirement for a safety harness or lifeline, but we were required to wear a life vest. It was explained to us that at this height, we would probably not survive the fall but the safety vest would make it easier to fish the body out of the water. Even I knew this was not the reason, but I did not let on, just played the dumb engineer and let them have their fun.

I handed one of them (there were about four people there watching) my hard hat so I would not lose it during the inspection. Question: How many construction workers does it require to watch a new engineer inspect a weld in a hard-to-reach, dangerous position that does not have to be inspected because it was already inspected in the fabrication shop before shipping to the field? Answer: Apparently it takes four people, or at least that's how many people were used.

I sat down on the ten-inch beam and swung beneath the beam while wrapping my legs and arms around the beam and scooted out to the subject weld just like I had crawled out on many a tree limb when growing up on the farm. I was in no hurry to complete my inspection but took enough time to study the weld good before climbing back to the edge the same way I climbed out, calmly retrieving my hard hat and declaring that that weld was sufficient and proper. I quietly turned around and left. I was never asked again to perform such unnecessary tasks for their amusement because of the way I handled it.

I felt that I performed my assigned tasks properly and represented my company in a professional manner but would have a lot more to observe and learn before I could consider myself a professional engineer. This was an excellent employer who treated me with respect and dignity, and I felt that I was giving them my best. It was a good relationship. I don't know of many companies that provide company cars and put so much trust and responsibility in the "new engineers" hired only for the summer job. I have to believe that some of my professors in school and especially the CE department head must have given me a strong recommendation for me to have secured the job.

I have learned later in life that success is often the result of what you do on previous jobs and assignments, and for a successful career, you need to have positive accomplishments and leave each client

ENGINEERING

happy with your performance. Not every assignment will be a success but if you work hard and do your best, the successful ones will outnumber the less successful and you can move forward in your career. I have another story about success related to this company that I will share with you in the next chapter in which I am interviewing for a long-term job.

CHAPTER 5

Return to School and More Education

After I completed my summer job, I returned to Mississippi State and enrolled in graduate school majoring in civil engineering with an emphasis in structural engineering. It was good to get back to familiar surroundings and people I knew. Even though I had been gone only three months, I felt like I had gained several years of experience and was ready to take on new challenges. Graduate school was a lot different, and the expectations were much higher for the students seeking an advanced degree.

I was able to secure a graduate assistantship for my time in graduate school. I received a monthly check from the school for teaching an undergraduate lab and grading papers, among other duties. I was selected to teach a hydraulics lab for three hours one day a week. I had a class of around fifteen to twenty students each semester that were required to perform and observe hydraulic experiments and write lab reports on each experiment. In the lab there was a Plexiglas channel mounted on a machine that could change the slope of the water flow and a pump to change the speed of the water that flowed through the channel. There was also a pipe system consisting of various sizes of copper pipe complete with valves to demonstrate flow through different-diameter closed-piping systems.

We covered such topics as friction flow in open channels, friction flow in closed pipes of varying diameters, hydraulic jumps in open channels, and laminar/turbulent flow in open channels. Most structural engineers do not deal with hydraulics on a constant basis. This was an excellent way to show flow around objects by using a dye inserted into the open-channel equipment that represented the flow of fluids, such as wind or water, around structural objects and the turbulence created by different shapes. If you are designing any type of structure, there is no way of avoiding wind loads on the exterior of the structure unless you are designing something for the space station or on another planet.

The program included a thesis in order to graduate, which required a lot of research and independent work. Initially, each graduate student had to select an adviser who would guide us through our class selections, research, and independent study. I chose a professor who had taught me structural analysis and structural steel design in my undergraduate classes. The graduate school typically lasted for two years with a fall and spring semester and occasionally some classes in summer school. My first semester consisted of only three courses, but they were so intense it seemed like six. Each class had a lot of homework and reading every day in addition to the periodic exams.

I enrolled in several advanced courses of subjects covered in my undergraduate classes such as Advanced Reinforced Concrete Design and Advanced Structural Analysis. The advanced concrete course was centered on the American Concrete Institute Building Code Requirements for Structural Concrete (ACI 318) and Commentary. We learned in detail not only how to design structural concrete but why there are certain requirements and how to use the ACI manual properly. This course enabled me to cover topics that had only been mentioned briefly in the initial course. Concrete design has always been a little harder for me because a structural member made of reinforced concrete is not a homogeneous material and depends on the composition, arrangement, and strength of the components to determine the capacity and performance of the member. In preparing the structural code, the authors spent a lot of time researching, testing, and developing the requirements to be used by the engineer.

The advanced structural design course was probably my favorite course in graduate school. Not only did it reinforce the concepts and methods of structural analysis but it also introduced additional methods of analysis. I took this course during the summer, and it was two hours long, five days a week. I developed a strict work/study regimen to keep up in the class. I woke up around 6:30 every morning, showered, got dressed, and went to the school cafeteria for breakfast, which was finished by eight a.m. I went to my dorm room and studied or worked on homework for two hours and went to the class from ten until twelve. I then ate lunch and went to a part-time job I had on campus, working in a maintenance department of a research lab from one until five.

At five o'clock, I either jogged or worked out for about forty-five minutes to get my exercise and then ate at the school cafeteria or went off campus to get a meal. I was usually back in my dorm room by eight p.m., and I studied and worked on homework until around one a.m. when I went to bed and slept until around 6:30 a.m., got up, and repeated the same schedule. Sometimes I would study until two or three if I was behind or had a test coming soon.

On the weekends, I would either study or catch up on homework if I was far behind. But most weekends, I would drive home Friday night (it was about a two-hour drive) and visit my parents and return on Sunday afternoon. On the trips home, I would study and work on homework when I was there so I could have a night off on my return to school on Sunday. This dedicated routine helped me to be a leader in the class and earned me an A for the course. I also was allowed to grade papers (for my professor) for the structural courses the following year as a part of my assistantship because of my understanding and grasp of structural analysis.

At the time I was in graduate school, no one had personal computers, laptops, or smartphones on which you could perform calculations. If you needed precise calculations, your choices were to submit IBM punch cards to the school computer center or use an expensive

ENGINEERING

handheld calculator. The IBM punch cards were made of thin cardboard roughly three-and-one-fourth inch by seven-and-three-eighth inch with one corner clipped off so the cards were oriented correctly. There was a machine similar to a typewriter (think keyboard with mechanical levers that struck a carbon tape and placed a single letter or symbol on the paper behind the tape for those of you who do not know about typewriters) that, when struck on the keypad, would punch a rectangular hole in the card as well as type the character struck on the top of the card. The location of the rectangular punch indicated to the computer the character being struck. Therefore, one IBM punch card represented one line of program instruction to the mainframe computer.

If you did not have enough time or access to punch the cards yourself, you could fill out a blank coded sheet form and some other student working in the main computer building would punch the cards for you and return the punched set of cards. If your program was long enough or had lots of data to enter (only one line of data per card), it could be several hundred cards long and you would see people carrying around boxes of IBM cards with their programs or data to and from the mainframe computer building. For large sets of data, you would often see an X or a symbol placed on the top of the IBM cards so you could tell at a glance if a card was out of line or missing.

For several of the hydraulic lab experiments, I provided a set of IBM cards and left them at the computer center marked specifically for the hydraulics lab. Each student was required to provide IBM punch cards with their lab data to be added to the computer program cards at the computer center and run their own program. The program cards were then left for the next student. The computer center had a card reader that would read the cards and run the programs in a matter of a few seconds.

By the time I was in graduate school, there were several students who purchased handheld calculators at a cost of around $600, which was a lot of money in those days. Therefore, not everyone could afford them. When most of the students showed up with calculators, the professors began to add more problems to the exams because they could be solved much faster than using a slide rule. This forced the students who did not have a calculator to either buy or borrow one for the exams. Only a few years before, I could remember taking many exams with my slide rule. And in a fifty-minute class period, you could only have time to solve three or four problems. The calculators were expensive but not very powerful in comparison to modern calculators or computers. Today you can buy a calculator for $10 to $25 that will do more than the $600 calculator that was available then.

Those students who did not purchase calculators still used their slide rules, and some people were very efficient at solving problems that way. Often groups of students would work together to study for exams, but most homework was based on individual effort and usually was submitted to the professor to check our work. Our capabilities were already known to our professors from our homework assignments and discussion in class, but the exams were given to satisfy the university requirements and to separate students who were very close in their abilities to comprehend and put into practice the lessons we were learning.

Other classes my adviser suggested for me were in areas that I did not have much experience and were difficult to understand. I remember a class in the theory of elasticity that introduced new concepts and ways to solve problems using series and advanced math. I understood the concepts and was able to solve the problems and pass the tests, but I believe that these methods were more useful for an engineer who was to pursue a doctoral degree, work in research, or teach at the college level.

I also was enrolled in a class labeled Advanced Calculus for Applications, which mostly consisted of math majors. I did find this class more applicable to problem-solving but I made the mistake of taking the class in summer school. At that time, the university had a summer school consisting of two semesters. The particular semester I chose to take the calculus class, the school was rearranging their schedule, and the entire class lasted only five weeks. To make matters worse, I had taken all my math courses in junior college three years earlier. I did not get my act in gear for the course and received an F on the first test. I received the first test score only one week before the final exam and the second test score (which was a C) the day before the final exam. We only had two regular tests, and even if the second test score was improved, I received a D in the course.

This was not acceptable for a graduate-level course, so I repeated the course at normal speed and received a B. The only two acceptable grades in graduate school were an A or a B. Even though I struggled in this class initially, I found it to be useful in solving some of the more involved problems. This course helped me to understand the background for a lot of higher math and also to understand how much of mathematics is derived from natural occurrences and the natural order of things.

I also learned a very valuable lesson. Not everything you try will be successful in the way you want it to be. Sometimes you have to regroup and try again. Learn from your mistakes, and try not to repeat them. Do not blame your perceived failure on something or someone else without looking inward to see if you could have done more to secure success. I will discover this again in a more dramatic way when I relate the details of my efforts to complete my thesis.

I also took a class in numerical analysis, which introduced me to additional methods that could be used to solve engineering problems as well as reinforce those methods I was already familiar with from previous courses. At the time I was in graduate school, a lot of structural analysis used methods that could be calculated by hand or use of a slide rule. This included moment distribution, slope deflection, analysis of trusses using summation of joint forces, and graphical analysis of trusses. Moving into the computer age, we were now learning to analyze structures using finite element analysis, matrix analysis, and combinations of analysis methods internal to the software programs being developed.

When I was in graduate school, we used the computer to solve multiple problems at a much faster and more accurate rate but still relied on our understanding of the interactions and stresses of the structural system to verify our solutions. One of my concerns for the structural engineers of today is that they will simply accept the solution presented by the engineering software as the absolute answer and not question whether it is a reasonable one. Sometimes the acceptance of the solution should be based upon experience or verification by another method if the engineer has limited experience. I believe this is a very important reason for engineers to have a number of years of experience beyond school before obtaining a professional engineering license. It has nothing to do with having the knowledge of how to analyze structures but more with having the experience in knowing how to interpret the analysis and determine if it is appropriate. In other words, sometimes "you don't know what you don't know."

I enrolled in a couple of soil mechanics classes—Advanced Soil Mechanics and Slopes and Stability—and received an in-depth knowledge of soil mechanics. I considered this a big plus since all structures are generally supported by a foundation system surrounded by either natural soil or added fill material. This has often given me additional knowledge and experience that many of my structural peers did not have. In fact, my adviser told me that if I enrolled in another soils course, I should consider getting a different adviser. My soils classes did provide me with an additional opportunity while in school. The head of the civil engineering department at Mississippi State was a soils engineer and had secured a job analyzing the soil for a lake levee to be installed near Tupelo, Mississippi. He was searching for a student to run the lab tests, and after inquiring from among the soils majors and not getting any takers, he asked me if I was interested.

I had just gone through an intensive course including laboratory work that included all the lab tests he needed. He instructed me to keep up with my time and said he would pay me for the work. The core samples were received from the boring site, placed in a secure moist atmosphere, and I began to run lab tests. These included grain-size analysis, moisture content, compaction tests, direct shear analysis, triaxial shear analysis, liquid limit, plastic limit, and consolidation tests. I also ran a slope stability program on the school mainframe computer. After completing all the work, I had spent forty hours that I reported to the department head. He asked how much I wanted, and I thought about asking for $2 per hour, which was the minimum wage at the time, but I did not. He suggested $5 per hour, and I was shocked but graciously accepted.

ENGINEERING

I found out later that he had his secretary type up the report using standard typical details and descriptions of the work performed along with the lab results and the slope stability analysis I provided. After reviewing and placing his PE stamp on the report, he issued it to the client and probably charged them $5,000. I did not feel bad after that for accepting $200 for doing probably 90 percent of the work.

There were several lessons learned here. Even though I was very careful in all the lab tests, they were reviewed by a professional engineer who had secured the work and assumed the responsibility for the report and deserved the benefit of the report. It was important for an experienced engineer to issue the report ensuring that the levee was constructed properly and the lives of the public were not placed in danger. It also encouraged me to continue my education, gain appropriate experience, and obtain a professional engineering license for myself. I realized that even though I might not get rich, engineering was a noble profession that would support me in future years.

During graduate school, I had enrolled in a class for similitude, which is a fancy way of saying scale modeling. It was an interesting class, but all the people in the class had a structural major and needed specific structural courses for graduation. We talked the professor into changing the course to Theory of Plates and Shells. I have never heard of any group of students changing the course they were taking, but we did it. It helped that there were only three students in the class, but I believe that being in a small class in graduate school helped enormously with the professor-student ratio and provided much more individual attention to the subject matter. This class helped me to expand my understanding of how the basic courses were used to solve more complicated structures.

For this course, our final was to provide a design for a thin-plate cylindrical shell roof. It took the first half of the course in order to understand the applied forces, the resulting stresses, and the analysis methods required to produce the design for a concrete cylindrical shell roof. The remainder of the course consisted of performing the actual design and producing the dimensions of the concrete shell roof complete with reinforcing details and design sketches for the construction of the roof. This opened my eyes to structures I had seen pictures before such as the Opera House in Sydney, Australia, and the Los Angles Museum of Art but had no idea how they were designed or constructed. Once again, I realized that many natural mathematical shapes were employed in the design such as hyperbolic paraboloids and catenaries. I was so engaged with the idea of Theory of Plates and Shells that I used it as a basis for the upcoming thesis that was required of me to complete graduate school.

Of the thirty semester hours required to graduate, twenty-four hours were classroom instruction and six hours were for thesis development and production. The thesis had to be an original work that had not been done before. After discussions with my adviser, I presented a proposal to develop a thin concrete shell roof with a geometry such that the weight of the roof would be supported by a roof structure that was always in compression under vertical dead and live loads. Therefore, it would only need reinforcement for wind loads and for temperature and shrinkage in the roof structure. The previous problems with thin-shell roofs were complicated by the need for extensive reinforcement to support the dead and live loads that, in turn, increased the thickness of the roof, which then required heavier reinforcement and so on.

For the first three semester hours, I performed research and developed a plan for the thesis. I was able to determine that many people had designed thin-shell roofs and that there was a lot of background material to be researched. No one had specifically attempted to define a new shape with the entire shell in compression under continuous loading that would require reinforcement only for wind loads and for temperature and shrinkage requirements. After the feasibility of the thesis was assured, there were multiple parts in accomplishing the actual research.

The first part was to construct a physical model to demonstrate the theory. The second part was to test the physical model to see if the model matched the theoretical assumptions. The third part was to develop equations that would define the shape and relative dimensions of an actual thin-shell roof that matched the physical model test results. The fourth and final part was to complete my thesis report including the theory, the developed equations to define the shape, construction of the physical model with pictures, test results performed on the physical model along with appropriate references and credits, and submit it for review and approval by the thesis committee as a requirement for completing my master's degree.

I eagerly began to think about ways to develop the model. It had to be large enough to demonstrate the theory but small enough to be able to handle. My first model was constructed outside of my parent's home in rural Mississippi. A base dimension two-feet wide by three-feet long was used, and a rectangular frame out of two-by-fours was placed on a set of wood sawhorses to elevate above the ground. A layer of thick polyethylene plastic was attached to each three-foot side, and the plastic sheet was draped between the two sides to form a flexible cradle form for the concrete shell. The ends of the plastic sheet along the two-foot side were unattached and free to move.

A thin layer of mixed cement sand and water with the consistency of mortar was applied to the plastic sheet to an approximate thickness of three-fourth inch. The cement mixture was mixed in a wheelbarrow and applied with a mason's trowel as uniformly as possible. The thin shell was left to air dry and cure approximately two weeks before carefully lowering it to the ground and turning the shell over. The plastic sheet and the frame were removed from the shell. It had imperfections due to the crude method of construction, but it stood on its own and appeared quite stable. I was encouraged by my first attempt but knew that it needed much improvement.

My next model had to be constructed at school so I could properly document and test the shell. It also needed to be a little larger to get reasonable results, and there had to be a way to handle the shell since, with increased size, it was getting much heavier. I utilized a frame inside of a frame in order to support the shell and also to be able to rotate the shell from the position of construction to the position of testing.

Once again, a two-by-four frame was constructed to support the shell, and it was place inside a larger rectangular two-by-four-braced frame. The inside frame was attached to the outside frame using half-inch threaded pipe with pipe caps at the centerline location along each side and three-eighth-inch diameter bolts at the four corners of the frame. This allowed the frame to be secured during construction, and by removing the four bolts at the corners, the entire inside frame with shell roof attached could be rotated 180 degrees longitudinally to go from the shell cast in tension on the plastic sheet to compression in a matter of seconds.

Construction of the concrete shell was similar to the first model, but extra care was made to control the thickness and uniformity of the shell. After the concrete shell model had cured for a minimum of twenty-eight days, two sets of piano hinges were epoxied to the edge of the shell roof and attached to the inner two-by-four frame. After the epoxy had fully set, the model was rotated with the curved shell facing upward and the plastic sheet was removed.

In order to test the stresses in the concrete shell roof model, the strain gauges were epoxied to the concrete shell at strategic points and initial readings were taken with the shell in the inverted tension position. The shell was rotated to the upright compression position, and the strain gauge readings were recorded again. It was difficult to get accurate readings due to the method of construction and the size and weight of the model tested. At one point, an oscilloscope was used to determine the readings and the overhead fluorescent lights in the lab had to be turned off due to interference.

The end result was that some strain gauge readings were in line with what was expected from the theory. Some strain gauges did not reflect any change from the tension to the compression orientation, and some of the readings were actually slightly opposite of what was expected. In retrospect, the quality of the type of construction and quality of the mix did not lend itself to accurate results. I believed the theory was sound and one day would be established, but my only course was to continue to improve on the physical model and hope that the results would be different the next time. There was nothing that suggested it would be any better the next time or the time after that.

I have found myself to be a determined person—some people even say stubborn—but if the war was to be won, this battle might need to be lost. This is the only time in my professional career that I have backed away from anything and prefer to say that a better path was chosen. After discussing with my adviser, a new thesis topic was undertaken, and my master's degree was completed two years after I left school. One of the reasons I left the campus before completing my degree was the fact that I was broke and needed to start work and earn some money.

The events of my first full-time job will be covered in the next chapter, but in order to complete my master's degree, the thesis work had to be completed. My new employer allowed me to work on my thesis using their computers and resources but only if the work was done on nights and weekends and did not interfere with my normal scheduled work activities. I coordinated with my adviser and the graduate school by transmitting information back and forth over the next two years and registered for my last three hours each semester and received an "Incomplete" each time until all requirements were met.

The first step was to submit a new proposal for the thesis. I had accepted a job with a petro-chemical company and chose a related topic for my thesis. I also chose this particular topic because it could be accomplished utilizing computer tools and known methods or methods that could be derived. My title for the thesis was *A Computer Solution for Circular Tank Foundations*. The chemical, petro-chemical, and refinery industries all used API-type circular tanks for storage of various chemicals and fluids.

My thesis proposal offered to provide a complete circular concrete tank foundation supported on piles with concentric circular beams and connecting concrete slabs between the beams. Input for the computer program would be the tank diameter, the weight of the tank, the volume and weight of the contents, the pile capacity, and the concrete strength. My proposal was accepted, and the research was begun to make sure it was a unique solution. I gathered all my references.

For this particular solution, references were needed for the solution of circular beams for any length, radius, and cross-sectional dimensions and for circular pie-shaped concrete slabs of varying radius, thickness, and angular dimensions. At the center of the foundation, a traditional circular pile cap was used for the first circular beam. The number of piles required was based on the total load divided by the pile capacity. The computer determined the number of circular beams and the pile spacing for each beam, rounding up to the next number of piles needed and rounding down on the pile spacing in each circular beam.

After solving for the number of circular beams and pile spacing in each, the pie-shaped segments were analyzed using vertical loads for shear, moment in two directions, and torsion. The loads from the connecting slabs along each side were placed on the circular beams, and they were analyzed for shear, circumferential moment, tangential moment, and torsion. The loads were then calculated for each pile and compared to the pile capacity. If the calculated loads exceeded the pile capacity, the number of piles in the circular beam in question was increased and the analysis was repeated until the pile capacity exceeded the loads placed on the piles.

Most of the theory behind the circular concrete slabs and beams was generic and could be used for specific cases and was not applicable for a wide range of variables. For the analysis, equations were developed using calculus to determine stresses for any number of variables such as radius of the structural member, angular dimensions, and dimensions of the cross section in order to provide shears, moments, and torsion about the structural members of the concrete foundation. Classical methods of analysis were used when available, and proper reference was noted.

The actual computer program was written in the Fortran computer language, which was short for *formula translation*. It was the accepted method and language of computer analysis at the time and was appropriate for the task. Most of the work was involved with taking the theory, accepted analysis methods, and newly developed equations and putting it into computer code and requesting output in a form that could be understood and used by the average engineer to provide a foundation for a circular tank.

In those days, Fortran did exactly what it said: it translated equations into computer code that could be used for analysis and produced output that could be understood. It did the same thing that an engineer could do by writing an equation on paper, filling in values for the known variables, and calculating using either a slide rule or an expensive handheld calculator. The results would be the same, even though the computer was much faster and probably more accurate if coded properly.

What Fortran did not do was to provide graphs, charts, shear, moment, and torsion diagrams. I am sure there was probably some software capable of doing these things or at least some computer or structural major was working on it for their thesis or dissertation, but it was not widely known nor was it necessary for the completion of my task. Fortran also did not provide an explanation of what was being done, so comment lines had to be added to explain what was being done at each step in the analysis.

This is still true today. Anyone who writes an original structural-analysis program or provides a lengthy set of data input must explain what is being done at each step of the process to be an effective tool in understanding the analysis. A great deal of the computer code was inserted to explain the process to the casual or first-time user of the program. Each of the utilized standard analysis equations as well as the developed equations specific for the program were input in the proper order. Instructions were added to the program to provide the desired output in the correct form with any warnings or failures along with the reason.

After the program was completed, several months were spent debugging the program, adding enhancements or comments, and testing the program for all types of input to test the reliability and the error messages when bad data was entered. At the time I had access to a time-share computer to do all my input and testing. It was all batch runs. The program was sent to St. Louis via telephone for analysis, and the results were sent to New Orleans for printing. The results were sent overnight and received the next day. Depending on the time of day the program was sent (I only worked nights and weekends), it was at least a two- or three-day turnaround sometimes to realize that you had a typing error and the run was no good or aborted. For those of you who were wondering why it took several months to complete this portion, now you have a clue. I was a full-time employee and a part-time student.

In order to submit the program, it had to be converted to a mainframe program. I had done all of my development using McAuto time-share in St. Louis and received permission from my employer to run the program on the company's mainframe computer, an IBM 360. Unfortunately, it did not like some of my internal coding, so several weeks were spent recoding the program and retesting. After successfully running the program on the company mainframe, I then had to return to the university and make sure it would run on their mainframe, which was a UNIVAC. Just like the previous computer, it did not quite like the internal language, so I rewrote it once again to satisfy the UNIVAC. After finally succeeding, my attention was then turned to the actual paper document that was required.

After completing the program and debugging and testing, the thesis had to be prepared and presented to the thesis committee for review and approval. The first part of the thesis consisted of presenting the need for and the solution of the design of a circular tank foundation. A detailed explanation of the theory including standard analysis techniques and the derivation of equations required for the solution of the foundation were presented. In order to represent the geometry and the varying forces and stresses on the individual structural members, inked plans, sections, and free-body diagrams were provided at the appropriate sequence in the text of the thesis. The full text of the computer solution was provided, and a sample run of the program with results was presented. It was finally complete, and now it was on to the oral review and exam by the thesis committee.

The thesis committee consisted of my adviser and two other professors from the civil engineering department. First, the thesis was presented and explained in detail, and they all had a chance to question what, how, why, etc. I felt confident about this since it had occupied a great deal of my time and effort over the past two years. Then they moved on to asking general questions related to structural engineering and somewhat-related topics that I did not feel as comfortable with. I could not tell if they were happy with my answers or disappointed because they appeared very grim and solemn in their

ENGINEERING

attitude. I suppose they were probably not happy to be there either, but I was the one who was under the stress. Regardless of what they thought, I was approved and was able to receive my master of science degree in civil engineering, majoring in structural engineering, in August of 1975.

I still have my deck of IBM computer cards with my thesis program on it and a printout from the program. I also have two copies of my bound thesis. Some people use their thesis for multiple uses. Mine was used for only two times that I am aware: the first was obviously to graduate, and the second has an interesting story behind it. I was at work one day (probably in the early eighties), and while on a coffee break, a coworker who was a mechanical engineer was complaining about a problem he had trying to find a solution for a curved beam.

He needed to analyze a curved beam for an uneven number of supports to incorporate into a silo design supported by load cells. He had previously been using a book of standard equations by Roark, which only had solutions for an even number of supports. I casually remarked that I had developed an equation for my thesis that would give results for any number of supports for a circular beam. He came by to see me later that day and asked if I really had such an equation, and I pulled it from the bookshelf and showed him the equation. He took it and tested it with an even number of supports using the Roark reference book and got the same results as he did while using my equation. He was convinced it was a good equation and used it to design the appropriate beam. There is a lime silo in Magnolia, Arkansas, whose support beam was designed based on the equation I derived in my thesis. It made me feel really good that my equation helped solve a real-world problem.

As previously stated, I was broke and needed a job to produce income. In anticipation of graduating, the process of interviewing had begun. I signed up for interviews with the companies visiting the campus and began to talk with anyone that had a need for a civil or structural engineer with an expected master's degree. Thirty-minute introductory discussions were made with various oil companies, petro-chemical companies, state highway departments, railroads, and the Army Corps of Engineers.

There was an opportunity to work for the local office of the Mississippi Department of Transportation at a location less than twenty miles from my parents' house with friends and cousins and low pay. Most of the oil companies such as Exxon, Texaco, Shell, etc. offered better pay but would require me to move to either New Orleans, Houston, or to points farther away and had a rather regimented order of training and company moves that I was not ready to commit to at this early stage of my career. The Army Corps of Engineers was looking for personnel to fill their Vicksburg, Mississippi, research center; and it was tempting until they indicated that I could be placed in a particular section to repeatedly design one type of structure over and over again. This was certainly a good way to become very specialized and knowledgeable in a particular area.

I am sure there would have been other opportunities to do other things, but in my inexperienced brain, all I could think of was designing the same concrete beam over and over thousands of times. It did not seem that appealing. This very thought has carried over with me through my career when I am interviewing others. I always make sure the applicant realizes that regardless of what position we are looking to fill, there are always other opportunities available if they are interested.

I was also very interested in a job with the Illinois Central Gulf Railroad based out of Chicago, Illinois. I was offered a site visit, and I flew to Chicago for more detailed interviews. This was my first trip on a commercial airline, and flying into and out of O'Hare Airport was quite an eye-opener. The flight into Chicago was very good. I almost missed the flight out of Chicago due to the tremendous traffic getting to the airport. I was on a bus traveling from downtown to the airport, and the trip started three hours before the flight was scheduled to leave. I made the flight only because the plane I was scheduled to fly on required service work in order to fly, and I was able to catch the flight even though it was late.

This was not my first flight in an airplane. I had started taking private flying lessons the week before and had taken control of the

ENGINEERING

Cessna 150 two-seated trainer from the flight instructor and flown the plane for a few minutes. The difference between the Cessna 150 and the Boeing 727 were quite dramatic. I thought the interview was great, and I was offered a job as a railroad inspector surveying bridges and tracks in an area from Shreveport, Louisiana, to Birmingham, Alabama, and from New Orleans, Louisiana, to Memphis, Tennessee. I was promised to be on the road five days a week and at home every weekend. It was an ideal job for a young man. But I had recently become engaged to be married, and I did not think it was the best choice to be on the road every week. So I reluctantly declined.

I also had an interview with the company I had worked for during the summer between undergraduate and graduate school at the university. When I walked in for the interview, I was greeted by the head of the offshore construction division engineering group, and we were both very familiar with each other. I had worked for a supervisor who reported to him as the head of the group, and I had interaction with him while I was there. We remembered each other, and it was a very pleasant greeting. He immediately stated that they were not looking for an engineer with a master's degree and could not pay as much as some other company who was seeking those qualifications. He also told me that he was aware of my capabilities and if I wanted to come to work for them on a full-time basis, I would be offered a job with no further interview necessary.

He then turned to me and asked if I would like to join him and the head of the civil engineering department for lunch, and I graciously accepted. I felt special being singled out to eat with the department head, and I had suspected that he had given me a good recommendation for the initial summer job. This department head was also the one I had provided the levee analysis for the soil report previously. I got along well with and respected him, and he also provided me with support and help in starting my career. Since I thought I would get a better offer from someone else, I did not pursue this guaranteed offer.

My interview with a petro-chemical company was the most appealing. The petro-chemical company had an office and manufacturing facility in Baton Rouge, Louisiana. They offered a career doing a variety of civil and structural engineering projects that would span the entire range and allow me to discover what I liked best but not get bored with doing the exact same thing all the time. I was invited to visit the office in Baton Rouge and drove down for the detailed interview.

I talked with several people and learned what I would be doing as a new civil engineer. They discussed how I would be working to modify structures and facilities for plant expansions and growth. I guess I was a little naive because I still envisioned working on brand-new plants and "greenfield sites." The majority of my work would end up being modifications and add-ons, but I would get to work on new sites also. The location in Baton Rouge was interesting because it was close enough to visit my parents with a four-hour drive but far enough away to be able to start a new life without feeling like I had never left home. They also agreed to letting me start work and finishing my thesis on my own time. It was the best opportunity, so I accepted a full-time job as a civil engineer working in the engineering department of the petro-chemical company located in Baton Rouge, Louisiana, on September 10, 1973.

CHAPTER 6

My First Real Job—What Does a Civil Engineer Do?

I was married on September 1, 1973, and began work for a petro-chemical company nine days later on September 10, 1973, at their downtown offices in Baton Rouge, Louisiana. We had located a furnished apartment in mid-city and drove into Baton Rouge pulling a small U-Haul trailer with all our possessions in tow. We were starting a new life with a new job in a new city and state and looking forward to what was ahead.

The first few days were full of paperwork and getting familiar with the new job. I was assigned an office that was shared with a more experienced civil engineer and had a desk with chair and a flat table to lay out drawings and reference materials. I also was able to share a bookcase for the few text references I had brought with me. It was easy to fall into the routine of arriving at work at 7:30 a.m. and leaving at 4:00 p.m. with a thirty-minute break for lunch. There were three other direct-hire civil engineers (one of which had been there for three months), a couple of contract civil engineers, and four or five contract designers with a civil engineering group leader to head the group.

The company had recently acquired a lot of work and needed engineers to staff the work. After I came to work, the company hired another half dozen engineers and added a couple more contract engineers in the next two to three years. It was a diverse group coming from Louisiana, Mississippi, Alabama, South Carolina, Virginia, and Michigan. I was given a job to install floor drains in an outside paved containment area that lasted for a couple of weeks and then was assigned to my first large project as a civil engineer.

Although this was primarily a petro-chemical company, they were very diverse in their holdings. The company was developed as a joint venture of a major oil company and a major automobile manufacturing company specifically to produce a compound to prevent knocking in gasoline. The product known as tetraethyl lead (TEL) was developed in the laboratory and was produced on a large scale at their facility in Baton Rouge, Louisiana, adjacent to the large oil refinery. The first major production facility was begun in 1937 and was completed by 1939. TEL was later phased out and is no longer available for use in motor fuels in the US. The phase out was partially due to the environmental concerns over lead contamination and partially due to the fact that the lead destroyed the catalytic converters that were now used to satisfy the environmental concerns over other emissions.

I will not attempt to defend or condemn the use of lead in gasoline, but for many years, the use of an antiknock compound made possible the use of gasolines in the emerging US market. Without it, there would have been an entirely different development. Many items banned in the US have been needed, but there are many others that have been proved later on to have been a poor decision. There are also many decisions made in the name of science that have no scientific basis.

The production of TEL provided many by-products that were used for other manufacturing processes, and the early company was basically a chemical company. The company was purchased by a paper company in 1962, and the original name was kept. Over the next few years, the company diversified into plastics, aluminum extrusion, life insurance, coal production, and other chemical companies. My first large project was working on a paper machine rebuild in Rumford, Maine.

The project was the rebuild of an existing paper production line at the existing paper production facility. The facility was very old, having started in the 1800s before there was electrical service to the facility. It was built on the banks of the Androscoggin River, and the plant's first source of energy was a waterwheel in the river. The particular production line that was being upgraded produced fine-quality coated paper sheets that were utilized in the pages of the *National Geographic* magazine as well as other fine publications.

The plant received wood pulp from various sites owned by the company, and the pulp was turned into a white slush containing the wood fibers in solution and placed into a giant vat known as a couch pit. The solution was applied to giant rollers with screened mesh, and the water was drawn off until the sheets could support their own weight. The continuous sheets were passed around large rollers heated with steam to dry them before the sheets were collected on a twelve-foot-wide roll approximately ten feet in diameter and shipped out on railcars to the clients.

Bear with me for those of you who have spent many years in the paper industry and could describe the paper process in very fine detail, for I am only giving the briefest of detail. The plant was much more complicated than I have described and involved a lot of intricate detail on the monitoring and controls of the paper line. For me, it was the challenges of the structural system supporting the paper machine that was most interesting.

The first thing that caught my attention was the drawings produced for the existing plant facilities. Even though we were working on only one paper machine line, there were multiple paper lines within the facility. And some of them shared resources and had overlapping functions. Being the young inexperienced engineer, I was impressed by a drawing of the architectural elevation of the side of the plant. It was a large thirty-six-inch by forty-four-inch full-sized drawing on linen paper in ink, and every brick was drawn in its entirety.

Today most designers and drafters would use a typical block detail to show the brick for a small portion of the elevation view, and it would be drawn in CADD. There was not a lot of detail on the drawing other than the brick, so it would probably take an experienced designer a couple of hours to complete. The original designer for this drawing could have probably spent several days on this one drawing alone. After I got past the awe of the inked drawing, I was hit with the fact that this structure had sections of steel that I was not familiar with, and I had never experienced cast-iron columns before. I was given a reference manual of steel sections used in the late 1800s and early 1900s.

As the existing drawings were reviewed, it was discovered that the concrete floors in the building were of arch construction between the steel beams. The reinforcing was connected at the beam webs and was exposed at the midpoint of the slab where the arch was at the thinnest section. I had not seen anything like this before. It took me back to a basic understanding of how the slab section supported a negative moment near the steel beams and how the positive moment at the middle of the slab was supported by the exposed reinforcing outside of the concrete slab.

Another surprise was waiting for me when I specified a minimum depth of a concrete foundation to be at four feet below grade due to the frost line. The construction forces in the field requested that the minimum depth be lowered, and after initially declining to do so, I was told that they were having to chip out and remove solid granite to replace with concrete. After understanding the reason, I readily agreed to adjust the minimum depth. Another lesson was learned. Not all hard and fast rules must be adhered to if there is a compelling reason to alter, especially if in the engineer's judgment, there is a good reason to change.

My first trip to the site was eye-opening to say the least. I learned that good engineering based on sound judgment and thoughtful deliberation is the best approach. Most of the things I had learned in school were simply the tools at my disposal and the ways to solve problems. As I would discover many times, the real key was to determine what the problem was before I tried to solve it. Formal education is great and is necessary to handle difficult situations, but having the knowledge and experience to be able to put the training to use was even more important.

I remember especially one morning that I was picked at for being an engineer and always rounding up and taking the next-largest size for member selection. After lunch, we received a notice from the equipment vendor that had placed a 100 percent impact load on the equipment supports, and I was practically begged to see if I could make the structure any stronger and more substantial. I decided at that point that I should always try to do the correct thing, allowing for expected increases in loads and neither overdesign or trim the design to the bare minimum in order to satisfy the wishes of others. This philosophy has supported me well over the years. I remember the saying, "To thine own self be true."

On this particular project, we were required to pour new foundations in and around existing foundations, and the shapes of the footings were often modified to take advantage of the space available. Some footings were square or rectangular, but most were Z-shaped, C-shaped, H-shaped, or other irregular shapes that required more analysis to check for shear, moments, and anchorage length of the reinforcing. One particular footing was bearing on granite and did not need a lot of bearing area, but the amount of reinforcing needed for the proper bond stress was causing a problem.

I cast a W6x15 steel beam into the concrete foundation in order to meet all the concrete checks and requirements. All the designs presented by me were checked by professional engineers, but it was reassuring to know that I was given the opportunity to put my education in practice and gain experience because most situations did not exactly follow the textbook examples.

The project allowed for numerous trips to the site to observe the field conditions, solve problems on the spot, and gain valuable experience. Without this field experience, I would not have developed into the engineer I became. I think it is extremely important for young engineers to visit the field and study the practical aspects of the work so they can better apply what they see to the next situation.

GERALD W. MAYES, PE, RETIRED

One of my early experiences where I initially missed the mark involved a structural connection at the end of a long line of steam dryers used to dry the paper. All the support bents under the steam dryers were connected together at a column height of about twenty feet, a column spacing of about twenty feet, and a length of well over one hundred feet. At the end, the last support was welded to the last set of columns. When the process was started up, the last connection was sheared in the horizontal plane, and I was called for help. My standard answer, not aware of the actual problem, was to reweld the support and start it up again.

After having the same problem upon start-up, I looked at the problem in more detail because something was wrong and I needed to find out what it was and correct the problem. At this point, I decided to go back to the textbook and look at the temperature expansion at the problem joint. After checking, I determined that the forces due to temperature expansion from the nominal temperature to the temperature of the heated steam drums was significant and the connection, as designed, could not withstand the force. I redesigned the connection with a slide joint that would support the load vertically but allow the horizontal movement due to the temperature change, and it worked just like it was supposed to work. I have never forgotten to check temperature loads on supports when required again.

I settled in to learning the process of providing engineering support for the civil and structural components of the plants my company was building in order to support the production of various chemicals and products. Our internal engineering department was divided into discipline and support groups all working together to produce the final product. In our case, the product was usually a set of engineering drawings and specifications that could be issued to a fabricator, an equipment provider, or a construction firm that would provide the necessary items and install or construct them at the chosen plant site for the completed plant. Our disciplines included process design, piping/mechanical design, electrical design, instrument design, and project management. Each of the design disciplines had dedicated drafters and designers to produce the drawings, and they sometimes overlapped such as the designers for process design and

ENGINEERING

piping design. We were supported by cost control, estimating, scheduling, purchasing, and administrative functions.

Since we were an internal engineering department, most of our work was received from the plants and operating divisions of the company. A request was made to provide an early estimate of the cost to produce a facility, and the department contained specialists whose job was to provide conceptual estimates of the cost and schedule for these requests. If a project was deemed viable, then a more detailed estimate along with supporting preliminary documents was provided. At this stage, the effort was mainly concentrated among the project management team and the process and piping discipline. It was their responsibility to provide preliminary plot plans of the plant along with proposed equipment arrangements of the facility based on the process flow diagrams (PFDs, which were provided by operations with support from process design) and mechanical flow diagrams (MFDs). Note that MFDs evolved over the years to P&IDs (piping and instrumentation diagrams) but served the same function.

The equipment arrangement drawings were a more detailed view of the equipment and facilities that indicated elevations of key components of the equipment and structures and the physical location of equipment and the relationship between the equipment and the structure. Equipment arrangements were shown in plan, elevation, and section views in order to show the arrangement. Preliminary specifications for major equipment were also developed at this time. Typically, the plot plan, equipment arrangements, and key equipment specifications were initially approved for design by the operating divisions who were the sponsors and financial supporters of the project before the design of the project could proceed.

At this point a kickoff meeting was scheduled, and representatives from all areas were invited. The meeting was usually run by the project management group (who had overall responsibility for the project) with the operations group (who was sponsoring the project) assisting in the meeting. All discipline groups who had involvement in the project as well as all engineering supporting groups such as purchasing, estimating, cost control, and administration were included.

Sometimes a representative from the construction group or the research-and-development pilot plant would be included. All participants in the meeting were provided with copies of the approved plot plan, equipment arrangements, major specifications, any preliminary quotes for equipment, preliminary schedule, and the approved scope of the project. Each discipline received the needed information in order to start the design of project for their area and expertise.

After the initial kickoff meeting, regular scheduled project meetings were held with the internal engineering groups that may or may not have included representatives from operations, construction, etc. The purpose of the regular meeting was to make sure every group was maintaining their individual schedule, to report progress, to update a needs list from other disciplines and other groups outside the meeting, and to indicate progress anticipated in the next few weeks, such as a two-week look-ahead schedule. The meetings were not used to solve problems unless they were minor and could be resolved with a simple question and answer but rather to identify potential problems that could be resolved outside the meeting.

For the civil/structural group, it was important to determine the size of the equipment so it could be placed or supported within the allotted space or elevation provided, the support lugs or attachment to the structure or foundation, and the weight of the equipment to be supported. It was also important to obtain preliminary estimates of the structure size (length, width, and height) in order to begin the preliminary design of foundations and structural members for support. Most of civil/structural design begins at the equipment or the top of the structure or pipe rack and ends at the final foundation that supports it all.

Construction of the civil/structural drawings, however, begins with the foundations and ends with the structure or support of the equipment. In other words, we have to know everything before we can design, but construction has to have the design complete for the foundations when they begin construction. This often led to pressure to produce a foundation before all load data was final. We learned that with proper planning and scheduling, a sound design could be

ENGINEERING

produced economically that would satisfy all requirements for the project.

One of the unique design responsibilities for the civil engineer was to determine the requirements to properly support buildings, structures, and equipment on the existing or modified soil structure in the area of the proposed plant site. We would contract out to a geotechnical firm the job of performing soil exploration and testing in the field as well as laboratory testing and analysis of the existing soil. This information was presented to the engineer in the form of a soils report that described the proposed project loads to the soil and the condition of the existing soil, listed methods of field and lab testing, and provided recommendations for the capacity of the soil to support the proposed loads. Recommendations were also provided for strengthening the soil, such as lime or cement stabilization, or for the removal and replacement with a structural fill material that would provide the correct support. For weak soils, there were recommendations for proper compaction and for soils that were expected to settle over time due to anticipated loads.

Many times, there were specific requests made to the geotechnical firm to provide data such as the existence of contaminated soil, the corrosive nature of the soil, the permeability of the soil, and the presence of materials that would be detrimental to proposed foundations, roadways, embankments, etc. The typical soil report described the site, gave proposed loadings, and recommended foundation types and allowable capacities for the proposed types of foundations. The geotechnical firms would also recommend that we use their services to follow the construction in the field for proper construction and compaction techniques. These firms were also contracted to act as third-party independent inspectors to verify the work was being performed according to specifications.

Most civil engineers have a general knowledge of soil testing and soil capacity (some are very experienced) but do not have access to the drilling equipment or labs in order to perform the tests and, therefore, rely on special firms that perform this service on a daily basis. I learned very quickly how to evaluate the soils reports and when to ask for clarification or request that additional tests or recommendations be provided.

In order to request a geotechnical report, an estimate of the proposed loads had to be provided. In order to determine what type and size of foundations would be required, a more detailed estimate of the loads was needed. In order to do this, we had to look at the overall site, the equipment arrangements, and any preliminary equipment sizes and loads.

We first looked at the major equipment drawings and specifications. The mechanical engineers generally provided an outline drawing of the equipment based on the process flow diagrams and the mechanical flow diagrams to go along with the specification for the equipment. These preliminary drawings indicated overall size such as diameter, length, and sometimes thickness for vessels, drums, tanks, etc. For exchangers and small columns, a typical piece of equipment was used from a previous project. For pumps, filters, etc., a typical cut sheet was used. A cut sheet was a typical drawing complete with a chart that provided this information for equipment based on flow rates or capacities.

Preliminary drawings also provided enough information for the civil engineer to determine the surface area and shape in order to develop the wind load on the equipment. In most cases, the dead load weight of the equipment could be determined from the preliminary drawing or from simple calculations, and the live load weight could be determined from the drawings or from the capacity and specific gravity of the equipment. If the specific gravity of the fluid contained in a vessel was less than water, the specific gravity of water was used to account for hydrotesting of the equipment.

Once the preliminary dead load, live load, and wind load were known, a preliminary foundation design could be performed. If the site was in an area of seismic activity, this would also be taken into consideration for determining the initial foundation design. These were often based on the conservative side since it would be easier to reduce the size and depth of a foundation when the actual loads were known rather than increasing them after everything has been determined.

ENGINEERING

It was also very important to determine the preliminary size and type of foundations planned beneath buildings and structures. Most buildings for functions such as administration, maintenance, warehousing, etc. could be easily determined due to past projects or based on anticipated loading due to the type of occupancy. Foundations for structures supporting equipment, both enclosed or open, varied due to the equipment or storage required at the various levels and would be based on the equipment loads and specified live loads.

A specialized foundation requirement for all chemical, petro-chemical, refineries, and many industrial and manufacturing sites is for pipe supports. This includes all supporting structures that handle piping, electrical and instrument conduit and cable tray, material handling systems, utilities, and supply lines both in and out of the facility. Foundations for these supports vary depending on the size, complexity, and type of soil. Unless the facility is built for capacity and cannot be expanded, increased, or modified in any way, it is a well-accepted fact that these pipe supports will be modified and added to for future changes.

It is very cost-effective to design pipe support foundations for an additional support level. The engineering cost is the same, and the construction costs is only slightly increased for extra materials in the foundation. This prevents future expansions from having to dig up the foundation and expand them to support the new loads. This is very costly and should be avoided if possible. Many companies, including mine, had as a general practice the design of foundations for an additional future level.

Most projects will not support the cost of future additions, but based on industry experience, it is often desirable to plan for future expansions. Most pipe supports in the industry are constructed of structural steel, and they can be added to by field welding. It is a general practice for pipe support columns to be extended a few inches above the highest connecting beam for future welding. It is also a general practice not to run future piping or conduit over the top of support columns because they may have to be moved to make room for future expansions.

To determine the weights and sizes of the structural members for buildings, foundations, and supports, past experience from other projects was used to get a starting point. Main beams and girders were checked individually to determine if they had the proper capacity to carry the anticipated loads. The main members beneath equipment lugs and supporting legs were first checked for shear capacity and then for moment capacity.

If the span was short, the shear usually controlled. If the span was long, the moment usually controlled. If the girder supporting the equipment was one of the main members of the wind-load resistance structure, then combined moment and axial load might control. All members were checked for unbraced lengths and compact sections to determine the allowable loads. It was important to know if the structural members were fireproofed and what type of fireproofing was to be used. If the member was to be encased in structural concrete, it added a lot to the weight and affected the capacity. In the final design, the structural members were all engineered and checked per the appropriate codes for compliance. But in the early stages, this gave the design engineer a quick check on a preliminary design that could be used to determine structure loads required for the preliminary design of the foundations.

Based on the engineer's experience, an early choice of the bracing system to be used for buildings and structures could be determined. The choices were usually moment connections between supporting girders and columns, X-bracing in selected bays, knee-bracing, diagonal bracing in selected bays, full-bent deep knee-braces, or a combination of bracing types. Each type of bracing has its advantages and disadvantages. If unhindered by other groups and disciplines, the average civil engineer would prefer X-bracing in all bays in both directions—therefore making the structure very resistant to deflections and sway due to wind loads and equipment forces but also not very usable because the X-bracing would interfere with equipment access, etc.

Most plant operation and maintenance personnel would prefer no bracing and every connection to be bolted for easy removal when the structural members got in their way. Obviously, we cannot do

ENGINEERING

this because the structure would be unsafe and not adhere to the proper codes for structures. Most piping, mechanical, electrical, and instrument engineers would prefer moment connections in all directions so there would be no interference with any of their equipment, conduit, piping, etc. This, however, is very expensive and not practical and should be used as an exception and not a rule.

What usually occurs and should happen is a compromise between moment connections and some type of bracing in selected bays based on the equipment arrangement and needs of the other disciplines. It is the responsibility of the civil or structural engineer to design the building with the proper restraints against deflection and sway based on the proper codes to accommodate the other disciplines, operations, and maintenance requirements. It is very important to choose the best bracing system at the preliminary design stage in order to make sure there is room for this system in the final design. A good preliminary design is also needed in order to determine the wind load path and the best estimate of loads to the foundation.

Once the preliminary design for the foundations and structures were complete, we began to develop better information as equipment quotes and vendor selection progressed. We were provided with equipment footprints with connection details, bolt sizes, and clearances required. We received pump and filter equipment anchor-bolt layout and sizes. We began to develop stair and ladder access to the various platforms and levels. We filled in support beams for grating between the main girders and equipment supports. We designed moment connections, baseplate details, bracing connections, etc. During the design, we continued to update our preliminary design when circumstances changed to match the changes and review for interference with other disciplines. At some point in the design process, we would usually develop a material takeoff of the completed design to check against the estimate design.

We received certified drawings from the vendors; performed our calculations and checks per the required codes, standards, and specifications; reviewed our design against other disciplines; and sent out approval drawings to the plant site for review by operations and maintenance. While waiting for the site review, we sent a formal intersquad check set of drawings to each discipline in the department for their review and approval. At this time, each discipline had a final check to see if we were all compatible with each other. And if there were areas in which we had a conflict or interference, it could be corrected.

Any comments received from the interdiscipline check or the site review were picked up and checked. The design package was then signed by the person responsible for drawing, designing, checking, coordinating, and approving the design drawing. These signed and approved drawings were issued for construction to the fabricators and the constructors at the site. Most of the time, the construction group was a part of the company but was a different organization, and we treated them as our construction client.

The instructions I received in school enabled me to prepare calculations, size structural members, and develop drawings with enough details to provide a fabricator or construction contractor with the information needed to complete their jobs. In addition to the drawings, I learned to provide specifications and standard details for the work to be accomplished. At first, standards and specifications prepared by others were used, and I gradually learned to develop and modify them to fit the particular project.

On projects that did not match previous efforts, a new set of standards and specifications were needed specifically for that project. Standards were used to eliminate the cost and time of preparing details from scratch and had already been proven on past projects. These details involved things such as concrete contraction, expansion, and construction joints; structural steel connections, baseplates, and gusset plates; roadway- and railroad-typical cross sections and details; schedules for baseplates, moment connections, pump foundations; pipe support details and schedules; and anything else that was used multiple times in order to represent to the fabricator or constructor what was required for the project. Most of these time- and cost-cutting measures were learned on the job and the experience built on the knowledge received in school.

In addition to the standards and specifications, other techniques were employed to complete the package. Plot plans or key plans were drawn to locate the work area within the entire plant site. Sometimes a small key plan was shown on the right-hand margin of the drawing to indicate what portion of the larger structure or foundation was shown in detail. Types of drawings were grouped together such as earthwork and site preparation, foundations and concrete at grade including dike walls and curbs, structural concrete and structural steel, roads and railroads, and architectural. In any given area, the plans were shown first followed by the elevations and sections and then the details. Any sections or details were cut or noted on the plans and elevations and referenced to the drawings that showed the details.

Certain information was always given for any project and was usually indicated on the first drawing, plot plan, or key plan. A set of general notes that covered the earthwork, foundations, concrete, structural steel, roadway and railroads, and architectural work was listed. The plot plan of the particular site was shown, and the specific work area was highlighted. Types of information given were the location and elevation of the benchmark used in the project; the coordinates of indicated horizontal control monuments (or references to existing locations within the site to be used for horizontal control); and the specifications for concrete, steel, concrete embeds, anchor bolts, reinforcing steel, etc. Also, notes that applied to multiple drawings would be given in the list of general notes. Each successive drawing would be referenced to the general notes listed on the first drawing. If notes applied to only a few drawings, sometimes the notes would be placed on the first drawing of that type, such as the architectural notes.

It was also very important that the drawings, details, standards, and specifications we were issuing to the field could be understood and built. Care was taken to ensure the drawings could be built and that we conveyed the proper message to the field or the fabricator. Things that could go wrong and sometimes did were specifying reinforcing steel and anchor bolts that overlapped in concrete pedestals, specifying and showing in detail certain field welding details that could not be made in the field without special equipment, and specifying the projection of an anchor bolt that did not provide enough length to go through the grout and baseplate with enough length to fully engage a nut with washer as specified.

We also had to be concerned with specifying reinforcing steel in a concrete section but not providing the necessary anchorage to fully develop the reinforcing per the ACI code and not specifying a proper bolt detail or weld detail that would handle the design forces we had calculated. These things were easy to draw but difficult to build, and it was our job to provide a good design that was functional, safe, and one that could be built with minimum effort. The engineer's job did not stop at the calculations but continued all the way through the issue of drawings and the successful fabrication and installation in the field.

On my first few projects, I followed the guidance of the more experienced engineers with whom I worked and gradually learned what to do, how to do it, and when to do it. The lessons I learned in those early days still apply now:

1. Do the best preliminary design possible with the information and knowledge you have, and keep it conservative.
2. Plan the preliminary design as if you will have to use it with respect to layout, bracing systems, and stair and ladder access.
3. Constantly update the design based on the latest available information.
4. Constantly check with other disciplines and groups to see if the design you are following will interfere with their design.
5. If you have a good idea, share it with others. It may be the best approach to the problem.
6. Be willing to accept the rejection of your idea if it causes other people problems.
7. Always be a team member and support the group effort.
8. Make sure your design and drawings are accurate, complete, and the best they can be.

ENGINEERING

9. Learn something from every project you do. Make sure you do not repeat any mistakes you have made. Make sure you improve on your effort on every project.
10. Be willing to help others to improve.

CHAPTER 7

On-the-Job Training

Most of the skills used in my career have been the ones picked up from others or learned on the job. This is important for any work or career, and engineering is no exception. School taught me the basic mechanics of the job, but training from others or by observation taught me the real who, what, when, and where to apply the mechanics for success.

There were formal classes provided by the company for critical path scheduling. We were taught the proper way to schedule a job considering manpower availability, equipment information, requested project completion or start-up in the field, and other factors related to the completion of our work. There were software programs running on mainframe computers such as Microsoft Project, which provided for the project schedule and was based on the required time and resources needed to complete individual tasks.

Each task has a set of previous tasks that must be completed before the current task can be started or completed. Sometimes a previous task must be completed a length of time after the current task is started or before the current task is completed. The software determines when tasks are started or completed due to the relationship of the associated tasks before and after. The length of time required to

ENGINEERING

do a task depends on other factors such as available manpower, the receipt of equipment information, the scheduling of manpower and equipment in the field, and sometimes, the seasonal weather.

This software also includes relationships between other disciplines so that scheduled dates and deliveries are realistic. If the schedule allows more time than required to complete a task due to other factors, then the schedule is said to have float (the time between the required time for a task and the time allowed to do the task without affecting any other tasks in the overall schedule). If a particular task or combination of tasks has zero float, it is said to be on the critical path.

I did notice that even though we were taught the proper way to do critical path scheduling, the projects I worked on often violated the principles of the method. According to the proper way to complete and update a critical path schedule, any changes (such as late equipment deliveries, delaying weather in the field, or a task that needed more time than allotted) may require the schedule to be changed to reflect these items and either reduce float time or change the end date of the schedule completion. It was acceptable to reduce float, but the end date almost never changed because of the commitment to the project.

What usually changed was the time allowed to complete a task such as design of a foundation even though the full time allotted was needed for a proper design. This would usually set up conflict between the project supervision who was trying to keep our commitments to the project and the design groups who were trying to provide a safe, operable, and proper design. These conflicts were not necessarily bad but a part of the process. The experienced project management team as well as the experienced design team were usually able to resolve the conflict to everyone's satisfaction, but the end date of the schedule rarely ever changed. The changes had to be absorbed, and the work continued. This practice sometimes caused the design groups to ask for more time and the project management group to request the work be done in less time because they knew problems would arise and everyone wanted to build in a cushion for the work tasks.

We were taught how to review vendor drawings for equipment and determine the location size and spacing required for anchor or connection details. The vendor drawings also gave valuable information related to the empty- and full-capacity weights of the various vessels and containers. We were shown how to determine if the proper wind load or seismic load was placed on the equipment and how it affected the connections. We also learned to review the size and location of nozzles, dip tubes, steam tubes, insulation supports, etc. for interference with structural supports, floor beams, and grating. For horizontal heat exchangers or long horizontal tanks with high temperature design loads, we learned how to evaluate the expansion of the equipment and how it affected our supports. We learned how to provide slide plates on the expansion ends of equipment supports. See the lesson learned with respect to this in chapter 6.

Coordination with other disciplines was vital to a successful project. The civil/structural/architectural group coordinated with the piping group for location, size, and extent of the main pipe racks, the individual pipe supports, and miscellaneous pipe supports. We worked with them to locate equipment within the structures and equipment on individual foundations and in tank farms to ensure the proper space for foundations, supports, piping, and equipment.

The mechanical group helped us with equipment details and information needed to design supports and foundations. The electrical group coordinated with us to provide grounding when the foundations were installed and provided information for electrical equipment. The electrical group as well as the instrument group coordinated with the civil group for cable tray routing and supports. The instrument group also provided information for instruments and controls requiring foundations and supports.

Occasionally, the instrument group required specialized supports for nuclear source and level detectors as well as load cell supports. The civil structural group as well as the piping group provided background location drawings for the other disciplines such as electrical and instrumentation. Each discipline was necessary for the other disciplines' coordination and success, and it was a combined effort that I learned how to work with the other groups and for the other groups when needed.

One of the specific tasks learned in the civil group was the checking of structural steel drawings and concrete reinforcing schedules. The concrete reinforcing schedules were usually provided by the rebar fabricator in order to verify the correct size, length, and correct bends for each reinforcing bar required for the project. As engineers, we often gave general instructions for reinforcing, but when presented with the reinforcing schedules, we had to review them based on all the ACI requirements such as lap splice length, lengths of standard hooks and ties, etc. We also saw reinforcing required for the installation of the concrete pour that we did not specify except as a general note callout to the constructor, such as the reinforcing bars required to properly support the top mat of reinforcing from the bottom mat of reinforcing in large pours.

The checking of structural steel erection and detail drawings were much more involved. This was something I had not seen before and was taught on the job how to perform. I learned that the drawings we issued for fabrication were used by a detailer to prepare fabrication drawings for the structural steel. Sometimes the detailer worked in the fabricator's shop, and sometimes the detailer worked as a subcontractor to the fabricator. In the early days, a sepia copy was made of our plans, elevations, and major sections, and a fabrication erection drawing was produced with a unique erection drawing number that labeled each steel member with a detail number. The detail number indicated on which detail drawing the member was shown and the number of the specific detail. The number of erection drawings could usually be determined from the number of major structural steel drawings issued. The number of detail drawings supporting each erection drawing varied from one or two up to as many as ten or more depending on the complexity.

Similar types of structural members were shown on the same detail drawing, such as columns or beams and girders. For example, column members that were similar utilized a detail drawn one time with notes added to the detail to indicate the various detail numbers and any differences required. Most details were not drawn to scale to take advantage of space requirements, and the dimensions noted were critical. There was more space needed to show connection holes and welds but not space required to show portions of the structural members that were not being modified in any way during fabrication. I am not sure why, other than to save space and time, but a dimension given in feet, inches, and fractions did not include foot or inch marks, as it was understood. All dimensions were given from a tail line, and you had to understand the line location and the distance from a column centerline or the centerline of a beam to that location. This was critical because every dimension was based on the location of that line. There were certain tolerances allowed for the steel connections, and the detailer generally used a tolerance when detailing a member that had to fit between two other members.

The detailer gave much more information to the fabricator than we provided on the construction drawings. The design drawings required adherence to the *AISC Steel Construction Manual* and the *Code of Standard Structural Practice for Buildings*. The detailer was required to follow the AISC code for bolt hole sizes and locations and lengths of bolts, nuts and washers. The detailer had to take into consideration the edge distance of connecting members, the clearance required to tighten the bolts using standard tools and wrenches, and the sequence of construction in order to safely install the structural members and connections in the field.

It was necessary to follow the AISC requirements for welding considering the weld size, type of weld, and welding procedures to be used. Tolerances were required to be included in the connections because of the tolerances allowed in the structural shapes used. A good detailer included all these factors and made it look easy, but an inexperienced or sloppy detailer that rushed and made mistakes could cause a lot of problems when trying to erect the steel in the field. Any issues caught and corrected at the detailer stage would pay for the effort in time and money. All the members, gussets, baseplates, splice plates, braces, welds, and bolts were listed on the detail sheets in a material table for accuracy in purchasing the materials and also for monitoring the quantities used and invoiced for the project.

ENGINEERING

In order to accurately check the erection and detail drawings, a color code was used. Highlighter pens were used to mark each item on the drawing: yellow for correct or accepted, green for removal, red for addition or correction, and blue or black ink for notes to the detailer. The drawing was complete when every item had been marked either yellow, green, or red with appropriate notes as needed. This included all references to other drawings and the correct quantity listed for each item in the material list. When all drawings were completely checked, a clean set of drawing prints were marked up showing only the green, red, and notes to the detailer with a stamp initialed by the checking engineer or designer, dated and noted as to the status. The status was either "approved," "approved as marked," or "correct and resubmit." This stamped set of erection and detail drawings were returned to the fabricator/detailer for correction.

Once all drawings were approved, the fabrication could begin for the structural steel. The time required for detailer drawings to be received after issuing drawings for construction or fabrication was usually from two to six weeks depending on the size of the project and the backlog in the detailer's shop. After approved erection and detail drawings were returned, the time required for delivery of the fabricated steel usually ranged from eight to fifteen weeks depending on the size of the project and the backlog in the fabrication shop. Often times the schedule of steel delivery was a deciding factor in which fabricator was awarded the contract due to the timing required for the overall project. On larger projects, the steel fabrication and delivery were broken down into smaller packages in order to receive the critical steel items first and avoid disruption of the overall project schedule.

The conflicting problem with the civil/structural design was always how to deliver the foundation design to the field and allow time for construction and curing of the concrete foundations prior to delivery of the structural steel. The structures generally had to be designed before the final loads and anchorage to the foundations could be determined, thus delaying the design of the foundations and the subsequent installation and curing of the concrete foundations. If the foundations could be completed immediately after the structural steel design was completed and sent to the fabricator, there was a window of opportunity to install and cure the foundations in the field while the steel was being fabricated. But it did not always work that way.

Foundations were often sent to the field based on preliminary load data to allow the field forces to mobilize and to order equipment and materials with just enough time to adjust the foundations if needed with a revision issued as soon as the final steel design was sent to the fabricator. If this could not be accomplished, then the structural steel was delivered to the site and placed in a lay-down yard waiting on the foundations. This was usually discouraged because management and operations viewed this as if we had not performed our jobs correctly even though many times it could not be helped. We did try to reduce or eliminate this, if possible, from the project schedule. When we did succeed in getting everything timed just right with no delays in the field, we took pride in knowing that we had performed our job in keeping the project on schedule.

We also learned how to apply the latest code revisions to our work. In school, working stress design for steel and concrete had been taught briefly, but we had learned to design concrete and steel using ultimate strength methods. Most of the experienced engineers at work were more familiar with working stress design, and we had to learn to apply the new methods to the accepted work practices. Most of the designs were made using slide rules or hand calculations, and it was easy to utilize the new codes.

When structural analysis programs were available, most of them used the current codes, so it was an easy transition. Other than the steel and concrete codes, most of the controlling regulations and design codes were updated and changed every few years, so we were constantly learning the new requirements. Wind loads, snow loads, and seismic loads were initially covered by the ANSI code, which was later incorporated into the ASCE-7 code for all loads on structures and buildings. In South Louisiana, snow loads and seismic loads rarely ever controlled the design, but the wind load was very important.

ENGINEERING

The first code I used to determine wind loads on structures was only thirty-seven pages long. Today that same code is contained in a two-volume set and is several hundred pages long. As a general rule of thumb, we would usually have a wind load of twenty-five pounds per square foot on structures and buildings one or two stories high and less than twenty-five feet in height. Today it requires a computer to sort through the various requirements of the wind loads to place on a structure to be in compliance with the building codes.

The results have not significantly changed, but the methods required to arrive at the same results have become much more complicated. A lot of this is a result of the review of wind-damaged structures and the use of the computer to perform more exact analysis. It is easy to understand why the results have remained unchanged. The wind loads are not any different today than they were centuries ago. We do have more complicated structures, and this often requires a more complicated method of analysis.

The early analysis was often done on individual members rather than the entire structure. Later on, entire structures could be analyzed in their entirety because of computers. The individual member analysis was not wrong, but it tended to lend itself to a more conservative result. As individual steel and concrete designs became more competitive, the use of computers to aid in the design became a crucial part of the engineer's toolbox.

CHAPTER 8

In-House Training versus Actual Performance

In all our company training, we were taught the correct way to handle projects. There was a procedure to give us directions for everything, which is an attribute for an established company. These procedures kept us in line with corporate directives and kept us out of trouble with management, the internal and external auditors, and our operations group for whom we were providing designs and drawings for plant expansions and modifications.

We had an establish procedure for filling out daily time sheets, which recorded our time against the various projects and their established charge codes and for time spent on proposals, preliminary estimates, and other charges not directly related to an approved project. This was necessary to separate capital charges from expensed charges and to keep our accounting books in proper order. We also had an established procedure for filing out and submitting expense reports to correctly place any valid expenses we had to the various projects and to overhead accounts. These were very straightforward requirements and usually did not cause any confusion to comply.

ENGINEERING

There were other project procedures that were established at the beginning of each project, and although they followed general guidelines, there were exceptions due to the nature of the projects. Depending on the funding of the project, some items were either expensed or capitalized and, depending on the state location, could affect the established charge codes and procedures. If the facility was being rebuilt under emergency conditions, such as a fire or explosion, it might have different funding or could have emergency funding. All these items affected the procedures and charge codes, so most every job was different. This led to time restraints for completing the project.

For a specialty chemical company, time was critical. If a project was developed in the research and development department, there was generally a time frame for which it was profitable. Some projects would completely pay for themselves in a couple of years, and other projects that were not producing product in a matter of a year or two were not worth doing at all. This would affect the project to the extent that a highly profitable product would be fast-tracked to get it to market as soon as possible, especially if it would only be profitable if it got to market by a specified time. The fast-tracked designs were often produced fast and conservatively because the increased cost of being conservative was not as important in the overall plan as it was to get it to market fast. If the product had a small but dependable profit, it would be looked at closer to try and reduce the cost and increase the rate of return.

I remember one project in particular that we rushed the design to get it completed early. The project was completed in the field, the plant was started up, and the product was produced. The product was stored in the small warehouse built for the project, and as soon as the warehouse was filled, the plant was shut down to never run again. By the time the product was ready for the market, the market no longer wanted the product at a price that made it profitable. So the production ceased. Examples like this were offset by others that paid for themselves in less than a year and, after that, continued to provide profit for the company. As long as there are more successes than failures and the successes produce an overall rate of return that is acceptable to the company management and stockholders, the process of looking for those high rate of return products keeps going.

Most projects started with a product that was developed in the R and D facilities. The initial production takes place in the laboratory on a workbench using small scaled-down equipment and containers and is called a bench-scale process. If it is successful, then the equipment and quantities are scaled up and the process is repeated in a small pilot plant facility. If the company sales team has done their work and identified a potential customer for the product, it could be produced for a short period of time to provide enough product for the customer to evaluate and test to see if it meets their requirements. These test runs are often called campaigns in which a specific product is produced in an established plant facility with full-sized equipment and may run for a full month in order to produce an appropriate amount of product.

If all has gone well at this point, the customer usually wants the specialty chemical company to produce a specified amount of product over a given period of time that will meet the customer specifications. Once a contract is signed with the customer, the operations group has a completion date for the project and has an estimated cost of the project in order to make it profitable. The process group at this time interacts with the operations group with help from other groups as needed to produce a process flow diagram (PFD) and begins work on a mechanical flow diagram (MFD—now called a P&ID for piping and instrumentation diagram) and a set of equipment arrangement drawings showing the proposed location, layout, and interaction between the major pieces of equipment, structures, and connecting pipe racks. Once this has been completed and approved, the project is ready for a kickoff meeting with all groups involved.

This seems like a simple process to follow and should be easy, but it is often difficult and trying to produce an economically feasible plant site that is safe and operable within the time constraints. As an engineer, it is our responsibility to see that the design of the plant does meet all the requirements of the project while following all the rules and requirements of the company, the applicable design codes, the requirements of the customer, the requirements of the operations

ENGINEERING

and construction groups, and any other difficulties that arise during the design and construction phase of the project,

For those of you who have heard of Murphy's Laws, there is one that states "If something can go wrong, it will go wrong." This is where actual performance comes into play. In the design meetings, we are always told that we will have vendor information available to do the job. Often vendor information is late or sometimes nonexistent. As a practicing engineer, you have to be able to design based on preliminary information and approximate loads and be prepared for surprises at the end of design that you have no time to correct. It helps to be clairvoyant and design based on what you think will happen rather than based on what you are told. Experience helps greatly for this, and the more experience you have, the more you can guess at the outcome.

Working for a petro-chemical company required a fast turnaround for designs and did not always follow accepted procedures. When a deviation was required, the proper approvals for the deviation were secured and determined to be in the best interest of the company while still following all the requirements to keep the company financially and ethically in compliance. I remember one project during development and design that required additional space for the equipment, and additional land had to be purchased from a neighbor to construct the structure and equipment. Another project required a technology that the company did not have at the initial project approval, and a small company had to be bought in order to use the technology needed.

It was not unusual for a project, while under design or construction, to still be developing the actual process to be used in the project. The plan and scope of a project often changed during the design in order to meet updated projected sales data or new developments in the marketplace. We were taught to follow the scope and schedule of a project but usually modified it during development to suit the changing needs. The company motto was "Changing to serve a changing world," but I had jokingly heard it referred to as "Changing for the sake of changing."

Many times, I was asked to change something to benefit another discipline or the client request, which did not provide a more favorable result for the civil/structural design. I was never asked to change something that would not work or something that was unsafe. In general, we first made sure a design was workable and would achieve the desired results and then, second, that it had the capacity and met the required codes and regulations governing the design. Third, it could be operated in a safe manner, and fourth, that it was as economical as practical.

If the design was not workable and did not provide the proper capacity or meet the required codes and regulations, then it was a nonstarter and had to be reworked. Next, if the design was not safe, it had to be modified or adjusted to make sure it was safe before finalizing. The last thing looked at was the cost. If the first three items could not be satisfied, then it did not matter how inexpensive the design was. It could not be used.

Once these three items were satisfied, we tried to make the design as economical as possible. We were trained to make a plan and then follow the plan, but most often, we continued to modify the plan as needed during the process to produce the best design possible.

CHAPTER 9

Handling Field Problems

Most engineering work is done in the office developing scopes of work, preparing specifications, determining the use of standard drawings and details, analyzing the structures and foundations, preparing estimates, and producing drawings that are issued for fabrication and construction. All this work precedes the actual construction in the field of the completed design. Sometimes issues arise while developing the design that must be addressed in the field before the complete design package is issued.

Underground obstructions must be considered in the design and often have to be located in the field and avoided if possible. Tie-in points for process, sanitary, and stormwater drains should be verified for coordinate location and elevation. Existing structures that will be connected to with new structures must be verified for location and elevation. Roads and railroads must be located and verified. Existing diked walls, levees, manholes, catch basins, fire monitors, load and unload spots, weigh stations, and pipe rack locations must be verified. For existing plants, the client will usually provide most of this information in the form of a plot plan or other existing drawings, but it should be checked and verified in the field before starting design. If other construction is in progress at the site that may affect the existing conditions, it should be checked as construction progresses to ensure there are no conflicts with the design in progress.

The largest portion of field problems generally occur during construction and can be handled with a request for information (RFI). RFIs can be used prior to construction to gain information about the site or conditions and may be used after construction during start-up if there is a need. RFIs can be used to gain information from any group associated with the project. Field problems most often occur during the underground or foundation phase of work. Neither the engineer or constructor can see what is underground until it is uncovered and have to rely on drawings, personnel experienced with the site, or other means (such as ground-penetrating radar) to determine what is below the ground.

It has been my experience that someone usually walks up, just after you have discovered an unknown item underground that is blocking the new work, and says they knew it was there but no one asked them about it. Regardless, the civil engineer is called to make a field trip, discover what is there, and propose a solution to remedy the problem so construction can continue. If the engineer has experience, they can often solve the problem in the field or carry the information back to the office for a quick check and issue a sketch to be followed by a revision to the drawing that corrects the problem.

While visiting the field for these type of problems, I have discovered the following:

1. Underground drain pipes that no one knew about
2. Underground drain pipes that were known but were not relayed to design
3. Broken concrete and brick rubble from previous construction
4. Abandoned wooden catch basins from an old plant
5. Rotting wooden decay from trees and bushes at the bottom of a foundation
6. Overpour from existing foundations that did not match drawings
7. Underground electrical duct banks not in the correct location per drawings

ENGINEERING

A lot of problems are related to connecting to existing structures, such as pipe racks, and existing equipment supports and open-air structures. The existing drawings can be checked and verified, but unless every member of the existing structure is checked and verified (this is a very time-consuming process and is often spot-checked to save time), some things are often missed. This is the case when the field verification does not know the exact connection point required from the design, resulting in piping and conduit interfering with connections and, therefore, discovered after the design has been essentially completed.

This type of problem can be corrected by moving either the piping, conduit, or structural member that will have the smallest effect on the overall project. These problems can sometimes be handled in a memo or picture taken of the problem and not involve a field trip. If the site was nearby and could be reached in a short time by vehicle, it was usually corrected after a field trip was made.

Some of the more interesting problems were not originally associated with a project but begin with a phone call from the field requesting an engineer visit the site and correct a problem that has been discovered by someone in the field. When equipment fails in an existing facility and it involves the removal or replacement of the equipment, structures and foundations may need to be modified for the replacement and support of the new equipment. Since equipment failure can happen at any time in the life cycle of a chemical plant, it is usually an emergency and immediate action is required. I will cover some of these replacements in a later chapter but will mention a few here that involved a visit to the field.

One of my earlier experiences involved the failure of a tanker truck frame that was discovered at a routine fuel stop in Nashville, Tennessee. The channel frame supporting the tank had split along the bottom flange and web and was only supported by the top flange of the channel. The tank was severely sagging at this point and subject to complete failure of the supports. It was an emergency, and I was asked to pack a bag, get a plane ticket, and get to Nashville as soon as possible. The only flight I could get was a multiple-stop route that started in Baton Rouge and made stops in Monroe, Louisiana;

Greenville, Mississippi; and landed in Memphis, Tennessee. There I changed planes and flew from Memphis to Jackson, Tennessee, with a continuation to Nashville.

After landing in Nashville, I rented a car and drove to the gas station on North I-65 where the limping tanker trailer was located. There were so many legs to the flights there was only time for drinks to be served, and there was no food provided. Luckily, I had carried my lunch that day, and every time we were airborne and served a drink, I could pull out a snack or sandwich and eat during the short thirty-minute flight. I had left on the first flight around 9:00 a.m. but did not get to the station until after 3:00 p.m. that day.

When I got to the site, an inspector from the company had arrived the night before, and he was waiting for me to tell them how to repair the tanker trailer so it could continue on the way. The tanker was filled with chemicals that, if spilled on the highway, would require the highway to be dug up and removed. I felt a little pressure but after analyzing the situation and determining what could be done, I acted like I solved problems like this every day and began to determine what should be done. The inspector had located a welder and had him lined up for the next day to repair the trailer.

The first problem was to pull the sides of the split back together. This was accomplished by lowering the jack on the trailer, which was used to support the trailer when detached from the truck but also served to raise the trailer and tank when still attached to the truck. I had determined that the trailer channel needed to be reinforced along the bottom flange and the web in order to safely continue the trip. The welder was not familiar with engineering drawings and sketches, but he knew how to weld. I asked him what size plate he was planning to use for the bottom flange and web connections, and he replied, "Whatever looks good." I began to quickly calculate what size plate would be needed to replace the bottom flange and web with a little extra to be conservative and decided what he was using was sufficient.

I again asked what size weld he was planning to use, and again he replied, "Whatever looks good." I calculated what size was required following all the steel code requirements and made sure that

ENGINEERING

he met those requirements. I had one additional request to make of the welder. I asked him to weld the splice plates along the horizontal direction and not along the vertical direction in which the steel channel had split. This was to ensure the stresses were spread out and were not concentrated along the same plane as the original failure. I could not tell the amount of stress already present along the plane of failure and did not want to utilize this area as the base for the repair. The welder did a very good job on the repair. I inspected and gave my approval, and the company inspector was happy with the quality of the work. The jack was raised, and the tanker truck was sent on its way to complete the delivery.

After the tanker trailer was back at the originating plant site, the inspection revealed that it had been modified and lengthened to carry a larger tank without any additional reinforcement or modification to the trailer frame other than increasing the length. The trailer was taken out of service until it could be properly repaired and inspected. We escaped a close call without any lasting consequences and experienced a lesson learned.

Most problems and issues can be traced back to a reason or what is known as a *root cause*. In this case, the framework was extended without proper modification and without proper checking by an engineer or responsible person who could have prevented the incident. I have usually found that problems are a direct result of someone doing something incorrectly. It could be the lack of proper information, use of incorrect materials of construction, improper or insufficient calculation of stresses and strains, improper or inadequate construction or fabrication, improper use or using outside the design capabilities, improper or inadequate inspection, and failure to remove from service those items that have become unserviceable due to corrosion, damage, or sometimes even "Acts of God."

Not all trips were that exciting. I was sent to a plant site to investigate a concrete block wall that was reported to be leaning and falling inside of a large warehouse. This warehouse was located in the middle of a cornfield in rural Indiana and was used for aluminum extrusion, storage, and painting. The eave height of the building was approximately forty-feet high, and the length of the building was separated by a concrete block wall installed by plant forces or a contractor directed by the plant. The wall had been installed to separate parts of the warehouse and different functions of the operation.

Upon arriving, I witnessed concrete blocks that had fallen from the top of the wall to the warehouse floor. The plant had barricaded the wall from any personnel for safety reasons. I was able to access the top of the wall from an adjacent mezzanine floor and was able to immediately determine what had caused the blocks to fall. From below, it was difficult to determine if the wall was leaning. From the top of the wall, I could tell that the wall was not leaning. For a properly designed free-standing wall to meet codes, intermediate beams or columns would be required. In this case, there was no load on the wall other than its own weight. It also had no wind load since it was entirely inside the warehouse. The supports for the warehouse were rigid frames that, under high wind loads, have a substantial movement at the top.

Since the warehouse was in an open cornfield with nothing to deflect the winds, the rigid frames were deflecting several inches at the top. When the block wall was installed, it was laid directly against the frame. In this case, the building was actually bumping the block wall and causing the blocks to loosen and fall. My suggested solution was to remove the blocks touching the frame at the top and install flexible insulation that would allow the frame to move and not knock the blocks from the wall and still separate the sections of the building. It was an easy solution but one that required a field trip to determine what was causing the problem before it could be solved.

The solution was easier than trying to make my flight back home out of the Indianapolis airport. It was mid-December, a snowstorm was coming, and I was on the interstate highway trying to get from the plant site to the airport. I was able to follow a snow plow and made it to the airport and caught the last plane out before they closed the airport. All in all, it was a successful trip. I came, I saw, I solved, and then I got the h— out of town. This trip helped me to realize that not everything is what it seems and all problems should be studied with an open mind.

ENGINEERING

Sometimes the problem was solved by other people. I have been sent on trips more than once where I listened to the preferred solution from someone at the site. They already knew how to solve the problem, but no one would listen to them. Since I was an expert (someone from more than fifty miles away), my opinion seemed to carry more weight than the local guy who worked in the plant every day. These local folks did not always have the education or formal training, but they had a lot of common sense and were well aware of what was wrong and how to fix it. This usually turned into a win-win situation. The problem was solved without a lot of effort on my part, and the local employees took pride in knowing what was required to fix the problem, especially if their solution was used by the "expert engineer" to solve the problem.

Not every problem was easily solved in the field, and many times the trip was only to gather information and pictures to be carried back to the office to study and analyze. A closer analysis would often reveal a problem area that had not been considered before and could result in a redesign effort to make the design operable and safe. Each time a new problem was solved, my experience level increased and I was asked to go on more trips to solve problems. Throughout my career the practice of learning from my problem-solving as well as learning from my mistakes increased my value to the company for which I was working. Each of us should practice this learning as we go through life in our careers, our vocations, and our family lives.

The other huge advantage of making field trips was to view the installation of a design that was sent out for construction on a drawing. If the engineer and designer only look at drawings and specifications, it cannot be fully comprehended because of scale factors and surrounding interferences. Designs take on a new meaning when they are viewed full-scale in the field. When designs are placed on drawings, they are scaled down to fit the drawing, and a six-inch-deep beam may appear similar to a sixteen-inch-deep beam. But they occupy different spaces and require different connections to install.

This is especially true if the steel beams are represented by a single line and the size designation only. Diagonal or X-braces are usually represented by a line with a short section drawn in at the middle of the member to represent the size and orientation but may not adequately represent the correct size of the gusset plate used to connect the members to columns and beams. The area at the connection of the main girders or beams to the columns is the location of the gusset plates connecting the bracing and is usually the favorite location for piping and conduit to be installed. Braces along the outer perimeters of open-air structures above the ground will require handrails. And these handrails, if installed per the drawings, may interfere with the structure bracing.

This is not always easy to see on a design drawing but is painfully obvious when the steel is erected at the site unless the clearances are taken into consideration during the design. The field is the perfect place to discover that pipes, conduit, cable tray, instrumentation, safety showers, etc. are blocking access or headroom to a maintenance area or walkway. The desired place to discover this is in the office before issuing the drawings but may escape the inexperienced engineer or designer.

The more times the site is reviewed during early construction, the more likely these issues can be resolved before the final construction design is issued. This is also a major reason for reviewing the discipline drawings before issuing them to resolve as many of these problems as possible. It is also advantageous to visit the site and observe overhead obstructions such as power lines, pipe supports, and installations that have been installed by plant forces or other projects since the start of the design. Sometimes one hand does not know what the other hand is doing.

One of my observations from my engineering career is the value of field experience in the successful completion of engineered design projects. When an engineer graduates from school with an engineering degree or a designer completes a technical degree, it is only the beginning. As an engineer, we are taught how to make complicated technical calculations and how to comply with the various codes and regulations. There is not sufficient time to study much of the practical aspects, and we only get an introduction.

While in school, we were presented with problems and expected to be able to perform a technical calculation to determine the proper

ENGINEERING

sizes of materials, the resulting amount of stress and strain under load, determine if the stresses were within acceptable limits, and if the structure or foundation was safe for its intended use. This is a good basis but does not address the many aspects of what is expected on the job.

The first major step is to determine what problem you are trying to solve. This holds true if it is an emergency call to the field or the normal design of a structure or foundation. In order to properly design a structure or foundation, you must determine what it is supporting and the associated loads. This includes dead load, live load, wind load, seismic load, operating load, maintenance loads, emergency shutdown loads, and construction loads. The governing codes, regulations, and client requirements must be met. Loads from other disciplines such as piping loads, cable tray loads, equipment loads, fork truck loads, etc. must be considered. Before beginning design, the geotechnical report, the site survey, existing structures and foundations, and tie-in points must be taken into consideration. The design must be coordinated with other disciplines and gain overall approval from the client. Field trips conducted during this initial phase of design are necessary to prevent conflicts and costly redesign near the end of the project.

Many of the field trips involved inspections or gathering of field data in preparation for the detailed design of an approved project. Much of this work, by its nature, had to be conducted in the field. We were often requested to provide an inspection of an existing structure or foundation in order to determine if it was adequate to support existing loads or additional loads anticipated. The typical project would begin by making a quick field trip to get a firsthand view of the structure to determine the complexity and extent of the inspection required. A set of existing drawings of the structure was also obtained, if available, to use as a base document for the accuracy of the drawings and the condition of the structural members.

For an extended inspection, coordination was required between the engineering inspectors and the plant operations and maintenance personnel in order to move freely through the site for the inspections by not disrupting operations or creating unsafe conditions. In most operating areas, the inspectors were required to check into the area control room to coordinate what each group was planning and make sure there were no interference issues. The operations had priority but were usually very accommodating in providing us access to conduct our work. Depending on the area to be inspected, it might require that a motorized manlift be used or scaffolding be erected to gain access. This had to be scheduled in advance and would often determine the date the inspections could start.

If assistance was required from the plant or construction personnel for ladders, excavation equipment, or crane with personnel baskets, it had to be scheduled. If any of the inspections required hot-work permits, they had to be obtained from the operations personnel in each shift in order to let them know what we were using and what we were doing. In some areas that were very sensitive to flammable materials and gases, using even a chipping hammer or any piece of equipment that could create a spark required the issue of a hot-work permit.

Personnel protective equipment (PPE) was required to be used while in the plant area. This usually consisted of long pants, long-sleeve shirts, hard hats, safety glasses, and steel-toed shoes as a minimum but might also include respirators, goggles, face shields, hearing protection, and gloves (all as appropriate for the use intended and the area requirements). Special requirements might include walkie-talkies, disposable coveralls, flame-resistant work outfits, or other safety equipment specified for the particular area.

If the particular area was designated as a confined space entry, additional permits were required, and the area had to be checked prior to entering for the presence of flammable gases and to ensure there was sufficient oxygen for personnel in the area. If dangerous chemicals were present or if there was not enough oxygen, those entering the area were required to have respirators and an oxygen supply with them to enter. Later, a requirement was developed for a preassigned person to keep watch at the entrance to the confined space in case of emergency and for a recovery system to be set up to extract personnel from the confined space before anyone entered. This usually occurred during construction while personnel were entering existing

ENGINEERING

tanks that were being cleaned or being modified in some way but, on rare occasions, did involve the inspection by an engineer.

Most structural inspections were marked on existing drawings depicting the steel beams, steel columns, concrete beams, concrete columns, concrete floor slabs, architectural walls, doors, windows, etc. and would indicate if the member was in good condition as installed, in need of minor repair (such as sandblast, prime and paint), in need of major repair (such as replacing or reinforcing sections of the steel beams and columns), or in need of replacement in-kind with a new member. Sometimes the replacement in-kind was not possible due to its location and the need for the plant to continue operations, and so a major repair would be performed in place to return the structural member to the original capacity or greater.

Concrete members were dealt with in the same manner. Minor repairs (such as sandblast and patch with grout) were handled on a scheduled maintenance type of repair whenever the structural members could be accessed. Major repairs were often required to be done during periods of operational shutdown to reduce the load on the members and sometimes required shoring to be installed. Major repairs might include chipping away some of the concrete cover down to the reinforcing bars and cleaning and coating the rebar to prevent further deterioration. In some instances, reinforcing bars might need to be doweled into the existing structural member to reinforce it properly.

After the reinforcing repairs were completed, the section would be grouted to return it to its former size and shape. Replacement of main concrete structural members was an involved process, and because of the connections and time involved for the concrete to cure and reach the full capacity, concrete members were often repaired or shored up with structural steel to save time and continue with the operation of the plant. If the damage was of sufficient extent, sometimes the layout of the facility was modified so the repairs could be done while in operation with the damaged members supported by temporary shoring. Then a short shutdown could allow the area to be supported by the new structural members.

Minor structural investigations were completed in the field, and sketches for repair were provided on the spot. Most, however, required more evaluation and determination of the best method of repair. If the structure was in very poor condition requiring multiple major repairs, it could be recommended that the process be temporarily relocated to a new area or halted until the members could be repaired or replaced with new structural members to support the ongoing operation.

A typical field trip for an inspection might be a few hours or up to several weeks. I was assigned, along with another engineer, to perform a structural inspection for a portion of a building while in operation. The inspection lasted for several weeks and consisted of inspecting and evaluating a two-story steel structure with concrete floors and a concrete roof slab with a flat built-up asphalt roof. The structure was approximately sixty feet by forty feet in plan and approximately thirty-six to thirty-eight feet in height. The bottom level of the structure was surrounded by a four-foot-high concrete wall, and the remaining sides were covered in corrugated asbestos siding. Access to the second floor was by way of a steel stairway or from the adjacent floor in the other part of the building.

The main purpose of the area was to take brine or salt water, concentrate it by evaporating the water, and dry the solution to produce salt, which was used for another part of the process. The structure was over twenty-five years old and had been subjected to continuous exposure to common salt (sodium chloride). The structural steel had corroded, and the concrete had been attacked by the salt on the reinforcing steel and the chloride ions on the cement in the concrete. Operations were continuing due to the need for salt in the process, but it was becoming unsafe to continue without a program of repair to the structure.

The structure was filled with large holding tanks, evaporators, and dryers that made access to the steel beams and concrete floor above the equipment very difficult. Scaffolding was required, and the construction forces at the site provided manpower and materials to provide proper access. We had good drawings of the existing steel and concrete members. The inspection was begun on the areas with ready access while the scaffold was being erected for the more difficult areas.

The two of us traveled to the site on a Sunday afternoon and were ready to start the inspection on Monday morning. We first met the operations personnel in the area and discussed what we would be doing and developed a good line of communications. We were dressed appropriately with long pants and long-sleeve shirts, hard hats, safety goggles, hard-toed safety shoes, a chipping hammer, a retractable tape measure, and a clipboard with our drawings and pencils/pens in hand to record what was found. It took a little bit of effort, but soon we had eased into a steady pattern of looking, chipping, measuring, evaluating, recording, and repeating as necessary in order to adequately assess the existing condition of the structural members and connections we observed. It was a very nasty job, but someone had to do it.

The operations produced a large amount of heat due to evaporation and drying of the salt, especially if you were on a scaffold a few feet above a dryer that was being fed by a natural gas torch. Think of a ten-foot-diameter hot-water heater with no insulation and no water turned on its side with a torch blowing through it at probably twenty-five to fifty times the heating capacity. We also had to wear gloves to protect our hands. If you touched the metal poles of the scaffold, it would burn your bare hands. The time spent on the scaffold was kept to a minimum, and breaks were required after only a short while above the heat.

Not all the inspection was directly over the heat source, and we moved along at a good pace. We inspected the steel columns, girders, and beams; the concrete walls, floors, and pedestals; and the concrete roof slab. We also inspected the steel stairs and steel girts supporting the siding. There was a problem accessing some of the floor area above major pieces of equipment, and we made a determination based on what we could gain access to and used our best judgment on the rest. There were advantages to our work conditions. When we stopped for lunch to go into the plant cafeteria, there was no need to add salt to anything we were eating. We just licked our lips and got all the salt we needed.

We were at the plant site about three weeks gathering information and then returned to the office to develop a plan for the repair of the structure. We evaluated the capacity of the concrete beams, pedestals, floors, and walls based on the level of deterioration. From this assessment, a determination was made if the members could be cleaned up and repaired or if they needed to be replaced. The same evaluation was made of all structural steel columns, girders, beams, and miscellaneous steel members.

If there was damage to a member in multiple places and the member could be replaced with ease, it was decided to replace the member as the best solution. If it was extremely difficult to replace a member, it was often repaired to bring it to the original capacity rather than replace. Some of the work could not be determined until the repair was begun and the true extent of the damage known. Our best judgment was used to determine the amount of cleaning, repairing, or replacing that would be needed.

Material takeoffs of all the required work and materials, based on the scope of work we provided, were given to the department estimators, who developed a cost of the repair work for the project. A project cost, schedule, and detailed scope of work was presented to upper management for approval. The project was approved and funded for repair during the next fiscal year.

CHAPTER 10

Sent to the Field—On My Own

The repair of the structure inspected in the previous chapter was to be accomplished by having a resident engineer in the field to supervise the repair and to make modifications as needed. A crew of construction workers from the on-site construction company was to be used for the repairs, and I was selected to be the resident engineer at the plant site. The project was expected to take approximately three months to complete. An overall plan was developed that indicated what was to be repaired and what was to be replaced to be supplemented by field sketches as the work progressed. The project was set up for Monday through Thursday, ten hours per day, with Fridays for overtime if needed to complete any repairs that were underway and had to be finished before leaving for the weekend.

Before this project, I had only been to the field for one-day trips and occasionally for two or three days. This was entirely different. The project was located on the ship channel near Houston, Texas, and there was an abundance of airline flights from Baton Rouge to Houston and back. It was determined that it was more economical to fly me round trip to Houston and back than it was to pay for a motel bill and meals over the weekend. Even though I was no longer in the office, I was able to be home most every weekend, and it was not that bad. Since we had a contract with a construction firm, it was also cheaper to have them provide a rental car by the month. I had the use of a car to and from the airport in Houston, to the motel, and to and from the plant site every day as well as a way to go out and eat every night.

I fell into a regular pattern of travel and work. On Sunday afternoon I took a cab to the Baton Rouge Airport and caught a flight out of Baton Rouge that carried me directly to the Houston Intercontinental Airport, picked up my luggage, loaded it into my rental car in the airport parking lot, and drove to my motel. Each morning I went to work using the rental car for transportation and returned to the motel each day. If there was no overtime required, I left the plant on Thursday afternoon after work and drove directly to the airport and dropped my rental car in the parking lot and caught the return flight to Baton Rouge. After arriving, I took a cab home for the weekend. The pattern was repeated in the following weeks.

To handle expenses, I was given an air travel card, which covered my airline tickets, as well as an advance of $200, which I had to reconcile each week on an expense report. Since I had not been transferred to the site, all my expenses (such as airline tickets, motel bills, meals, and miscellaneous expenses) were paid by the company. I used the air travel card to purchase tickets each week for the following week so I always had a ticket for the new flight. I stayed in the same motel each week and became familiar with the hosts and the rooms. On a couple of occasions, I requested a different room than I was given because the TV did not work well or some other reason I remembered about the room, and they were happy to comply. There was not a lot of extra things to do on the trips except eat out, watch TV, and sometimes go to a sporting event at my cost. The whole purpose of the trips was to repair the structure, so that was where my time was spent.

A package of drawings that indicated repairs to be made were provided to the construction crew, and as each problem area was uncovered, we determined if the repair was sufficient or if additional repairs were required. In many cases, there was no way to determine what was needed until the members were cleaned or removed. I had

ENGINEERING

a steady job following each discovery and was expected to provide a timely response to each repair that was uncovered. It soon became evident that the repairs would be much more extensive than originally proposed, and most of the work was handled with field sketches and verbal directions. Since the purpose was to maintain the capacity of the existing structure, no new analysis was required other than to ensure that the repaired or replaced members and connections had a capacity equal to or greater than the existing design.

It was much easier to repair on the conservative side than to analyze the structure to try to save a few pounds of material. It was also beneficial to keep the time and labor cost of repair and replacement in mind when making repairs, which usually was more expensive than the cost of materials used for the repair. Either way, I was tasked to guide the repair of the structure while it was still in operation, keeping the cost as low as reasonable and performing all of it in a safe and efficient manner. For a young engineer, it was a golden opportunity to learn and carry a lot of responsibility.

The construction crew consisted of a general foreman and leads in the areas of rigging, equipment operators, welders, and fitters along with general construction workers and helpers. Even though I was technically in charge of the group, I learned to go through channels and respect the ability of the people doing the work. As a young engineer in the field, I was tested by the construction crew as to my ability and my sense of humor. Construction workers often issued challenges or bets to the new guy to see how well he fit into the group. I was determined to show my sense of humor as well as bonding with the people I was depending on to make the repairs.

I was challenged to hold a bag of sandblast sand over my head for sixty seconds. I accepted, and the bag was lifted over my head. To my back, someone conveniently cut a hole in the bag so the sand would spill out and run down the back of my shirt. Even though I realized what was happening, I kept the bag over my head for the allotted time and met the challenge without complaining. As the sand poured down my back, the bag was actually getting lighter. So once I realized I had sand down my shirt, the task got easier.

The second challenge was a bet that I could not drive a railroad spike into the ground with one swing of a sledgehammer with my eyes blindfolded. I accepted, and we proceeded. The trick was to lay the victim's hard hat over the railroad spike and have the person strike their hat instead of the spike. At this point, they were a little afraid of what I might do if my hard hat was busted, so I was only allowed the swing at the spike. It would have been difficult to explain to management why I needed a new hard hat. I did, however, manage to drive the spike into the ground.

There were other minor challenges, but once my initiation was over and I was accepted, we became a close group that kept each other's safety and interest in consideration. I helped them in any way that I could, and they were one of the hardest-working crews I had ever seen even until this day.

To be able to be near the site, I was able to get a small construction trailer located about thirty feet from the entrance to the area. The trailer furniture consisted of two small metal desks, two desk chairs, two straight-backed chairs, two flat folding tables, and a couple of five-drawer filing cabinets. The furniture was equally shared between myself and the general foreman for the construction crew. All the paperwork, files, drawings, etc. for the project were kept there. All safety meetings and project meetings were held in the trailer. If I needed copies of any sketches or correspondence, I went to the main construction coordinator's office and used the reproduction equipment.

The crew size varied from time to time but usually was around eight to ten people. That crew handled the demolition, cleaning, concrete formwork and reinforcing, structural steel replacement and repair, crane operation and air come-along controls, removal and replacement of minor equipment, and wiring and piping during minor shutdowns.

One of the early tasks was to remove the siding on the lower floor in order to remove or repair the structural steel girts that spanned from column to column. The siding was corrugated asbestos and was typical of the siding materials installed on industrial buildings of the 1950s due to its excellent fire resistance and durability. This

ENGINEERING

was before asbestos was classified as a hazardous material, and there were no specific procedures for handling or removing materials containing asbestos. I also remember the house I grew up in was covered with asbestos siding, which was very common. To the best of my knowledge, the exposure to the asbestos siding has not endangered me or caused me to develop mesothelioma, but it did injure me at the construction site.

Some of the siding had been removed, and I was standing on the ground floor with my elbow leaning on a steel girt explaining to the crew and showing them what had to be removed when a large piece of asbestos siding broke loose and struck me. I was wearing all the required PPE, and the siding struck me in the hard hat, glanced off, hit my elbow (which was resting on the steel girt), and knocked me to the ground. I looked down and could see blood on my glove, and I immediately pulled the glove off, wiggled my fingers, and saw that my hand was not hurt. The damage was to my elbow.

I was carried to the local medical clinic, and an x-ray determined that I had a cracked bone at my left elbow. By this time, the elbow had swelled up considerably, and the doctor proceeded to squeeze the blood out of the elbow, getting it all over the floor and the walls. After he had returned the elbow to normal size, he put a Band-Aid on the elbow and proceeded to wrap my arm in a cast and provide me with a sling around my neck to support the arm in a folded position. The accident happened at around 1:30 or so, and I was back at work around 4:00 or 4:30 to finish out the day. I think it happened on a Monday or Tuesday, and I finished the week out and flew back to Baton Rouge on the weekend.

It was a little difficult dragging my luggage through the airport, but I managed to get it done. I had to wear the cast for several weeks until the bone healed, but I still continued to do my job and adjusted whenever necessary to get the work done. I think some of the younger folks today look for an excuse for not doing their work rather than an adjustment or alternative way of getting their work done. Life is not always fair or easy, but it is possible if we have enough drive and determination to succeed.

The work started along the outside edge of the structure, which was easily accessible. After the siding was removed, the steel girts supporting the siding were repaired and painted or they were removed and replaced. The girts were supported from the building columns, and the columns had to be repaired or have sections replaced before the new girts could be added. It was a slow progress of chipping, cutting, sandblasting, welding, painting, and sometimes replacing portions of the concrete wall or pedestals. If sections or beams or columns had to be replaced, the loads had to be supported by temporary shoring in order to keep the plant operating. It was my job to design and inspect the temporary shoring as well as provide the design, connections, and details for the partial replacement or repairs.

Every day was a new challenge, and I tried to stay ahead of the crew in order for the work to continue without interruption. When a new problem arose, I would provide a quick verbal solution for the shoring and the repaired or replaced parts so the materials and equipment could be readily available for the crew. While they were gathering the necessary items, I would analyze the shoring and repairs to make sure it was adequate for the loads and safe to use and make any adjustments needed before the shoring was complete and the damaged structural members were removed.

One of the major repairs was to replace the bottom section of a W12x65 column in the middle of the area that supported heavy equipment on the second floor. The column was damaged at the connection of the column to the baseplate. Both flanges were reduced in cross-section area by at least one-third, the column web was corroded completely through, and the baseplate was heavily corroded. I made the decision to replace the bottom five feet of the column with a new W12x65 and a new baseplate. In order to do this, the existing column had to be temporarily shored to support the loads from the four beams supported by the column. After shoring, the bottom of the column was to be cut out with a torch and a new section was to be welded with full penetration welds.

For the shoring, I used four standard weight eight-inch pipe columns. The height was approximately eighteen feet, and each column had a capacity of around 135 kips (1 kip = 1,000 lbs.). The W12x65

ENGINEERING

column had a capacity of around 311 kips. Therefore 4(135) = 540 > 311 with a safety factor of about 1.7. Unfortunately, this was the thinking of an inexperienced engineer. I did not estimate the actual weights of the operating equipment but assumed the existing column had the same connection conditions as the shoring columns and used the textbook capacities to determine if the four pipe columns were sufficient to support the load.

Once the shoring columns were in place, the bottom section was cut with a torch at five feet above the base. The existing column continued to support the loads until the very last section of the column was cut free. At that point, the existing column top dropped about five-eighths of an inch. This was true pucker time. We quickly checked all the shoring to see if any of it had slipped. It was all secure.

The shoring columns were placed under the connecting beams approximately four feet from the column because of space limitations, and this added to the displacement of the beams at the column connection (this was a classic case of a load at the end point of a cantilever). The sudden load at the base of the shoring columns caused a settlement of the floor paving. We quickly checked the equipment on the second floor, and everything was running with no problem. We monitored the equipment for a couple of days with no observed problems.

The columns were on twenty-foot centers, and the deflection limit criteria of L/360 would be 12(20)/360 = 0.67 inches. Our deflection was approximately 0.625 inches, which was less than L/360. After no problems were found, the new column section was field welded with full penetration welds then x-rayed to verify the welds, and we continued with the remaining work. This was a great learning experience, and I have never had a problem with shoring since.

I learned other valuable lessons. We cleaned the steel members by sandblasting. This location was on the Houston Ship Channel and was subjected to moist winds. Adding this to the salt environment of the process and we began to get instant corrosive attack on the unprotected steel within twenty-four hours. Our procedure changed, and we began to put a primer coat of paint on anything that was sandblasted during the same shift to prevent initial corrosion. Because of the demands of the process, we changed some of the concrete specifications in order to pour a higher-strength mix that would achieve an initial early set. This was used in areas that the equipment had to be taken out of service for short periods in order to make the repairs. It was a great learning experience, and I was learning it in real time.

One of the tasks was to repair a large girder that supported the second floor and various pieces of heavy equipment. A portion of the girder was damaged sufficiently that replacement was determined to be the best option. I think it was a W30 beam approximately twenty feet long. After properly shoring the beams supported by the girder, the old steel member was cut out and removed in pieces. There was no room for any equipment in the area to lift the steel members. Air tuggers (construction come-alongs using air pressure to lift loads) along with pulleys and cables were used to remove and install the new girder. This was accomplished with an air compressor and a lot of rigging skills that the crew proved to be capable.

It was slow and tedious work. Scaffolding was erected to be able to access the area and make the cuts and welds required. When the new girder was brought into the area, a fork truck was used to move the girder into position for the rigging to lift into place. The girder was balanced on the forks and was almost twenty feet long. The fork truck had to go through a door that was around eighteen- to nineteen-feet wide.

We had borrowed the fork truck from the plant maintenance, and they supplied the operator. When the fork truck operator slowly moved into position to pass through the building opening, he realized that the girder was longer than the building opening, and after several attempts, he said it would not go through the opening. While in school, I learned about Einstein's theory of relativity. One of the simple examples of the theory was that if you had a barn with open ends and a beam or member that was slightly longer than the barn, you could pass the beam through the barn. And if you did it fast enough, you could have the entire beam inside the barn at the exact same moment.

ENGINEERING

We could not do it that fast, but I had a very sharp general foreman who had been in construction for several years and had a few tricks up his sleeves (so to speak). He asked the maintenance operator if he could try and was given permission. Instead of approaching the opening straight on, he came into the opening from an angle and was able to get the right side of the girder through the opening before the left side of the girder was at the opening. He then turned the fork truck to the right, which allowed the right end of the girder to pass behind the framed opening and the left end to clear the opening as he moved the girder with the fork truck into the building. He did it all in one smooth motion and made it look easy. After successfully getting the girder into the building, the maintenance operator turned to me and said "How he do dat?" I knew exactly what he was doing, but the maintenance operator did not have a clue.

I continued working on the project, and after nine months (I was originally scheduled for only three months), I was replaced by another engineer from the design office who continued the work. This was one of the most valuable experiences as a young engineer that helped shape my further career. Many of the current new crop of engineers coming out of school do not get the experience of being able to see designs and repairs take place in the field. It is very important to be able to see what you are specifying and placing on a drawing brought to a successful conclusion. The field gives you a sense of proportion and highlights the ease or difficulty of installation and connections of structural concrete and steel members.

It is very easy to specify a field weld on a connection but often difficult or extremely difficult to install it in the field. It becomes even more difficult when trying to do retrofit work in an existing structure or on an existing foundation. When working in existing structures, it becomes very important to be able to physically place the member in the correct location due to steel flange interferences, other connections, piping and conduit, and other obstructions not always seen on drawings while in the office. I do believe the office atmosphere is necessary for high efficiency and production due to many factors, but the experience from the field is needed for accuracy and good design planning.

This was my first extended field experience, and I learned a lot about how construction is managed in the field. I was able to observe some of the designs I had produced in the office and saw successes as well as failures. I will discuss some of those failures in the next chapter.

For those of you who want to know what happened to the building, here is the history. The engineer who replaced me stayed for an additional nine months, turned the project over to a third engineer who worked on it for six months, and finally, a third-party engineer was brought in for a few additional months before the project was shut down. The reason the facility had to be maintained was to continue production for a short period of time. Had the repairs not been made, it could have caused a building failure and an abrupt halt of production. The repairs, although costly, allowed the structure to continue safely until the overall production plans called for a halt.

About one year after the repairs were halted, the production was stopped, the building was torn down, and a new plant totally different in scope and production was designed and built at the original site. This is a process that is fairly common to specialty chemical companies because most products have a short life cycle and are constantly being replaced with newer products and newer technology. I discovered that I would be involved in many facilities' changes and structures and buildings that would be repurposed to meet current market needs.

CHAPTER 11

My Inexperience Shows

While in the field on the building repair covered in the last chapter, I was able to observe what actually happens in the field, and I discovered that not every design placed on a drawing and issued for construction will work in the field. Sometimes it is due to poor design, and sometimes it is because the contractor does not install what is shown on the drawings. This can be due to poor representation of the design details and drawings, or it can be because the installer did not install it correctly due to lack of knowledge or experience. There are certainly a lot of reasons and excuses to go around. I will try to give a few examples of what could go wrong, what did go wrong, what was done to correct, and what should have been done to prevent.

I have already related my failure to properly design the beam connection that had high thermal loads on my first major project. It only took me three tries to get it right, and other than a little bit of embarrassment, the problem was corrected. I had studied the effects of thermal stresses and expansion. But it took a mishap for me to realize that thermal loads and stresses are real, and I have never made that error again. I have also related the problem I had with the column section replacement during the field building repair. Once again, I was faced with a problem that could have been solved if I had thought through the process fully before starting the repair. I can only say it was inexperience.

Unfortunately, the best teacher is to face your failures and learn from them. I am convinced that the reason engineers are not licensed to practice until they have at least four years of experience is to give them time to learn from their mistakes and realize that actual field conditions and problems are different from textbook examples. Knowing what the differences are and how to handle them is extremely valuable to the young engineer.

I was also able to observe a project under construction at the same plant that I had performed some of the design. One instance particularly stood out for me. I had designed a ring wall foundation for an API tank that used the existing soil in the area for compaction inside the wall. There was a requirement for using the existing soil that required compaction in lifts to obtain the proper soil density. Most of the site had been compacted using a sheepsfoot roller, and I observed the field contractor lifting a small dozer inside the ring wall to compact the soil. I thought it was odd that they were using a dozer but did not connect the dots. Dozers have wide tracks for support on soft earth and are one of the worst things you could use for getting good compaction.

About a month later, after the tank was constructed on the foundation, it was hydrotested in the field to make sure there were no leaks. When the water was added to the tank, the sides started to buckle. The soil beneath the tank inside the ring walls was not properly compacted, and the weight of the water compressed the soil and allowed the tank bottom to settle, which caused the sides to buckle. The remedy was to drill through the concrete ring wall and pressure inject grout into the soil beneath the tank. As the grout was injected, the tank bottom was raised and the sides were returned to their correct shape. My mistake on this was not to properly note the required compaction of the soil on the drawing to prevent the contractor from taking a shortcut on compaction and not stopping the work in the field when I observed the installation.

ENGINEERING

I also remember a structure I was designing that required a short column for an access platform to be installed on top of a steel beam. I placed the column member in the analysis software and ran the program. The section I had chosen was adequate for the imposed loads, and deflections were within limits. The design was approved, placed on the drawing, and issued for construction. Several months later, I received a call from the field saying that my column had too much horizontal deflection and I needed to look at it and suggest a repair. When I looked at the column in the field, I knew what was wrong. I had not modeled the member properly.

It was placed in the software with a fixed base, but in reality, it was on the top of a wide flange beam that was not capable of handling the torsional load from the column on top of it without reinforcement. I could walk up to it and place my hand on the column at about five feet above the base and shake the steel column with a small amount of force. The connection at the top of the beam flange had to be reinforced with gusset plates and connected to the adjacent beam for lateral stability. The best solution for this problem would have been to use an extended baseplate that bolted to the beam flange and to a connecting beam at ninety degrees with gusset plates in the web of the support beam.

I have learned that torsion is not my friend, especially if I use wide flange beams. I have learned over the years that it is usually easier to eliminate torsion rather than to design for it, and that practice has served me well. I have also discovered that just because the analysis software says it is okay, it may not be. In addition, I have also discovered that when the software says something will not work, it may be because the correct boundary conditions or releases have not been used. This is why it is so important to understand what is happening and not just blindly trust the computer answer as always correct. Computers are great. They can make fast and difficult calculations, but they need to be used in conjunction with common sense and engineering judgment.

On more than one occasion I started a design based on preliminary information, drawings that were not updated, or an assumption of what the field conditions were at the proposed site. If the project was tight on schedule and short on budget, assumptions were made to get the process started. When the project was well into the design phase and a field trip was required, things were discovered in the field that could have eliminated the need to change the design or, in some cases, start over. It is always best if the site is visited and interferences and problems are discovered before starting design, and I am convinced of this more now than earlier in my career.

I have found underground obstructions by way of the field contractor calling and asking me to come to the field and tell them what to do with the pipes, cables, foundations, etc. that were dug up. If I had thoroughly researched the existing files and viewed the site in the field, I could have discovered a lot of obstructions before they were dug from the ground. This is not unusual, and most of the time, it is simply viewed as a normal occurrence and part of the job. I have come to realize that much of it could be eliminated by being proactive and looking for these things before the start of design if possible.

While it would be great to always study the site in detail before starting, it is not always practical. The companies I worked for were in a competitive business, and time and money were valuable resources not to be wasted. It was always a decision of being extra careful and spending time and money or taking some risks (schedule and budget risks but not safety) and correcting any problems that were encountered. It would have been nice to have a crystal ball, but the closest thing to it was the experience I was gaining and knowing when to be cautious and when to use my engineering judgment.

On one particular project, I was designing foundations for a large manufacturing facility, and the geotechnical report had recommended using augur cast piles for the foundations. Geotechnical firms are among the most conservative group of civil engineers I know and for good reason. We ask them to drill and take samples at selected spots at designated depths and then analyze those samples and give us a recommendation for the entire site. The safety factors used are generally based on the amount of settlement that is expected to occur, and if the settlement is significant, it is considered a soil or foundation failure. Any respectable geotechnical firm would not be in business very long if all their recommendations resulted in excessive settlements or failures. Therefore, their recommendations are based on the lab results for the samples taken with an appropriate safety factor to prevent failure.

One of the ways geotechnical engineers reduce the uncertainty of their recommendations is through the use of pile tests. Pile tests are usually conducted at the site by installing multiple piles around a test pile and jacking the test pile against multiple resisting piles and determining the point of failure of the test pile. By using predetermined deflection and settlement criteria and plotting against applied loads, the ultimate capacity of the test pile can be determined. With this knowledge, the geotechnical firm can more accurately determine the safe load capacity of the pile.

On one particular project, the design was transferred to our group after the construction of fill material at the site had begun. The design was at least two months behind schedule when we started. Even though the geotechnical firm had recommended a test pile program be utilized, there was no time in the schedule to wait for the results. This is similar to asking the bank that holds your mortgage if you should add more insurance paid by you and due to the bank if there is a problem. I made a decision to go with the recommended pile capacities from the geotechnical report and forgo the pile testing procedure. The client hired an engineer to observe and review the design work we were doing, and it was in his best interest to question the design and justify his cost to the project. I had no problem with another pair of eyes looking at the design effort, but I was also willing to challenge any concerns he had about the design. The consultant engineer reviewed the geotechnical report and questioned why we were not using a test pile program. He was told there was not enough time in the schedule for the test piles and that I had decided to use the recommended values given in the report.

The consultant convinced the client that I was either designing the piles with too little safety factor resulting in a possible failure of the foundations or I was designing the piles with too much safety factor resulting in a waste of materials and time. Either way I was judged at fault for not using test piles. The design and construction of the site continued, and the client set up a test pile program to determine a more accurate capacity of the piles. The results indicated that the safety factor used in the design of the piles was appropriate and did not need to be increased nor decreased. Neither the client or the consultant brought up the subject again.

Some of my decisions and designs were not necessarily failures but maybe not the best designs either. As engineers, we strive to constantly improve our work and give the best we possibly can. We learn from others, we train, we seek feedback, and we share our success and shortcomings. Under current practice in most states, we are required to constantly take classes and learn new methods and refresh ourselves on engineering practices and procedures.

I am sure there have been other mistakes I have made over the years, but they were usually caught by others before they were issued for construction. One of the best ways to perform at the top of our skill level is to always have someone else review and evaluate our work. Even the young engineer can review and comment on the experienced engineer's work. They may catch something that was missed, and it is a good opportunity for them to observe how an experienced engineer performs a design. Everyone should be able to comment on any design for improvement and advancement of quality work.

CHAPTER 12

Gaining Experience on Additional Field Assignments

By gaining experience and being exposed to various projects in the home office, I was more valuable to send on special assignments. These assignments involved reviewing sites for future projects, performing site surveys and data-gathering trips, investigating field problems and providing solutions, conducting structural investigations and providing reports to clients, and other field-related issues. I was sent as a team member for field assignments and as a single engineer for items of lesser scope and duration. I took on all requests and handled them the same way.

As much information as possible was gathered prior to the trip. Information was gathered at the site such as existing drawings, survey information, field sketches, discussions with plant personnel, pictures of the site conditions, the initial determination of the problem, and the cause(s) and possible solutions. At times, the problems could be solved on-site with verification, field sketches, and instructions on the way to proceed. If it was an initial trip for a particular project, the information was gathered and taken back to the home office to aid in the design of the project. All field trips were different, but they all had common characteristics.

A typical problem was the installation of piping on top of a structural column that had to be extended for future use. The plant operations and maintenance groups usually ran minor amounts of pipe (not a part of a major project) in the easiest location, which was along the top of the structural column when the pipe rack was full. When major project plans call for the extension of the pipe rack to add another full level, the pipe rack columns have to be extended to support the new level. This is usually discovered during the first field survey of the proposed extension, and the assigned civil/structural engineer must decide the best method to proceed.

The preferred method is to relocate the pipe and extend the steel column with a full penetration weld to maintain the full integrity of the steel column section. Plant operations may not allow this modification because the piping is critical to the plant operation and can only be moved by taking a major shutdown and rework. This is generally frowned upon because of the loss of plant production, both in time and money. The structural engineer is tasked with leaving the piping in place and extending the steel column with another type of connection.

The column can also be extended by welding a new column section to the outside flange of the existing column and extending it vertically to support the new piping level. This is a less preferred method but has been used many times. Sometimes the column supporting the upper piping level is a stub column supported by the lower piping level beam that is inside of the pipe rack column. There are several issues with this approach. The stub column blocks any future pipe runs in this area of the rack. It may require the beams above and below to be reinforced with gusset plates at the connections to the stub column, which will add significant cost and time to the installation. Beams may be extended to the outside of the main pipe rack columns and supported by knee braces to transfer the loads to the original pipe rack columns. If the addition of the new piping level is larger than the existing rack, a new pipe rack—complete with new columns and foundation—may be installed on each side of the

ENGINEERING

pipe rack at the supports and braced back to the existing rack. I have designed each of these cases multiple times.

The best method is to design pipe racks with sufficient future capacity in the original supports to support new piping levels. Structural columns should be extended above the level of the existing pipe rack by six to nine inches (depending on the size of the rack and the structural columns) to allow for ease of future connections. It is also a good design practice to provide some addition space in the rack for future piping because there will always be a need for additional pipe space in the rack. If the operations and maintenance groups are provided a reasonable location, they will add any additional pipe in these spaces rather than over the tops of the main pipe rack columns. The best design is one that provides the support required in a consistent manner that is safe and operable for the plant forces. Trying to cut corners and put up something to barely get by that will have to be reworked in the future is not the desired plan of action. The job of the structural engineer is to provide a proper design that meets all requirements and one that can be defended on its merits.

The company I worked for had many plants and processes that were in the range of thirty to fifty years old and had begun to show age. Some of the structures had been repurposed several times for different products, and some had been producing the same products since the plant had been built. Because of the chemicals in the various plants, the structural steel and reinforced concrete members had deteriorated and needed repair to continue to be functional. Many of our projects were developed to maintain the structures or determine if they had to be repaired or replaced. These projects, much like the project in chapter 9 and 10, had to be inspected, a scope of work developed, a plan of action determined, and the time and resources allotted to implement. Once a project had been developed and approved, the inspection team was assigned and the work in the field begun. Some of these projects lasted for a few weeks and many continued for many months.

A production building consisting of concrete columns, beams, floors, and roof slabs had begun to show deterioration and needed to be inspected for repair or replacement. The concrete members had been installed over forty years previously and had come under attack from chemicals, mainly chlorine. The chlorine intrusion into the concrete had attacked the cement paste and corroded the reinforcing steel. Drawings were gathered, a thorough inspection schedule was developed, and the inspection of the structural members began. Chipping hammers were used to remove any loose materials, to determine the depth of deterioration, and to determine if the reinforcing steel was corroded.

Most of the damage was in the cement paste surrounding the reinforcing steel, and it was determined that it could be removed by sandblasting the member and replacing the damaged areas with grout. If the reinforcing was corroded, it had to be sandblasted to remove all corrosion and rust scale and was either coated to inhibit the corrosion or strengthened by adding additional reinforcing steel before grouting the structural member to return to original capacity. Very few members required removing and replacing in the column and beam sections. The concrete floor exhibited much more damage, and sections were removed and replaced in kind.

The real problem was the concrete roof. It was covered by asphaltic built-up roofing material but supported by a six-inch-thick reinforced concrete slab and reinforced concrete beams and girders. The beams and girders could be repaired and regrouted, but the concrete roof slab had sufficient deterioration that rendered it unable to support the roof loads. After studying it for several weeks, a solution was found that allowed the roof slab to support the loads without causing safety issues from falling chunks of concrete. This solution also prevented the roof from having to be removed and the production facility from being exposed to the outside elements.

A third-party firm specializing in fiberglass moldings was contracted to design sets of rectangular fiberglass panels with curved surfaces to catch the loose falling chunks of concrete that fit beneath the existing concrete slab between the beams and girders. These panels were bolted through the existing roof slab and anchored above the existing slab in a new six-inch-thick poured reinforced concrete roof slab. The new slab supported the roof loads and served as an anchor to support the fiberglass panels below the existing slab, which contained any concrete that fell from the existing slab. It was a sealed concrete/fiberglass sandwich panel that performed its job as designed.

This was a structure that was nearing the end of its life cycle, and the repair kept it going right up to the end. The design phase had been completed, the fabrication of the fiberglass panels was complete, the installation of the new roof slab was complete, and the installation of the fiberglass panels were within two panels of being complete when the project was shut down. Operations in the facility were halted, and the structure was torn down a few years later. This was an excellent example of inventive engineering and how problems can be solved if we allow ourselves to use our knowledge and experience to reach beyond the usual or normal way of doing things and try new ways to solve a problem.

Working for a company that designed plants producing chemical compounds and products involved a certain amount of risk. It has been shown that the average worker is safer at his job in a chemical plant than he is at his home, largely due to the safety practices strictly enforced by the plants and the continuous emphasis on safety. Whenever an accident occurs, it often produces spectacular results for the evening news broadcast and is portrayed as a result of some dangerous practice or policy. There is an old saying that goes "If it bleeds, it leads," and it describes what most reporters are looking to convey to the public.

It is true that accidents do happen, and they do injure employees, sometimes fatally. But what the public does not always see is the employees that control the accident, preventing further loss or injury to the plant facilities and other employees. What the public often sees is a picture of a large steam cloud rising (which is usually nothing but hot, moist air rising) or a giant fireball (which is usually the dangerous chemical being burnt off rather than expose the plant employees or the community to the danger). Having expressed my opinion that chemical plants are safer than most homes, I will relay to you some of my field experiences related to damage from fire, explosion, and wind damage.

One of our plant sites experienced an explosion during the start-up procedure of the unit turnaround that caused extensive damage. A pressure vessel exploded and destroyed everything within a twenty-five-foot radius. The concrete block walls of a nearby control room were blown into the control room, and much of the piping, instrumentation, and electrical supply was demolished. The explosion occurred on a Sunday afternoon, and I was sent as a member of an engineering team to repair the damage on Tuesday. We viewed the site on Tuesday and had a meeting with plant management on Wednesday morning to discuss the path forward. We were located in a construction trailer brought to the site and set up approximately one hundred feet from the center of the explosion. The construction trailer had a bathroom, but it was not connected. So we were supplied with a Port-A-Jon. In the meeting with plant management, we were given authority to spend and commit money as required to rebuild the unit.

Before I left the office on this assignment, I was not able to buy an ink pen on my authority, and now I was able to commit to thousands of dollars for the repair-and-rebuild effort. It was an abrupt change to say the least. It was extremely important for the rebuild to begin as soon as possible so it could be reported that construction had begun. I was specifically told to find a part of the unit that could be rebuilt, locate at least three firms, secure competitive bids, select a successful bidder, and begin work by Friday. That gave me parts of three days and two nights to get it done, so I started immediately.

I found a drawing of a concrete block shed that housed a hose reel used for fighting fires that had been damaged during the explosion. It was not critical to the restart of the unit, but it fit my purpose perfectly. I secured three names of contractors and asked them to the site on Thursday, gave them a copy of the drawing, and asked them to give me a price to repair or rebuild the block shed. I also stipulated that the work had to begin by Friday. I received the bids, selected a contractor, and the contractor began the demolition of the existing damaged shed on Friday. I was able to report that construction had begun on the rebuild.

I was part of a team that included a project engineer, a mechanical/piping engineer, an electrical/instrumentation engineer, and a civil/structural engineer. We coordinated all our work with each

ENGINEERING

other and had a meeting with the plant management every morning at the site at 5:30 a.m. Most days we worked until seven or eight p.m. before returning to the local motel for the night and starting all over the next day. The first order of business was to remove the damaged equipment, piping, electrical wiring, instrumentation, and concrete and structural steel. We were able to use the existing drawings to identify the items to be removed and the points where field connections were to be made.

The structural steel was marked up on the existing drawings and sent out for fabrication. The field steel connection details and concrete repairs were provided on field sketches. The structural steel required fireproofing, and it was added in the shop allowing for touch-ups and repairs in the field at connections. Once the structural members were identified to be repaired or replaced, it became a job of watching the construction crews install the structural members and provide assistance when needed and additional sketches and details as the need arose. At one time, the cost of the repairs was determined to be in the order of $100,000 per day, and this was in the early 1980s.

This assignment lasted almost two months, and I was glad to get back to the dull eight-hour workday in the office. This experience taught me to think on my feet and make decisions under high pressure and very tight time restraints. Some of my choices were conservative and may have been a little more expensive than I would have made under less strict conditions, but they were the correct decisions. I would not have changed any of them if I had to do them over again. I was congratulated on a job well done, and I felt like I had made a real contribution to the company.

One of the projects I was sent to the field was a fast-track project. This simply meant that it was a project that had a very tight schedule and the design was completed in the field with a team of engineers and designers using hand sketches (drawings produced in the field) while providing direction to the construction team building the project. This particular team consisted of a project engineer who was already assigned full-time to the plant and a group of piping, mechanical, electrical/instrument, and civil/structural engineers and designers assigned to the field for the short duration of the project. I was assigned to the field project for a total of seven weeks to complete the civil/structural effort.

At the first meeting of the group, it was pointed out that someone needed to develop an equipment arrangement of the plant addition, but the piping/mechanical engineers and designers (who usually did the arrangements) were busy working on mechanical flow diagrams and specifications for the equipment. I was designated as the person on the team with the least amount of immediate work required and was assigned the tasks of laying out the equipment location and elevation for the project. I worked as a part of the team to do whatever was required to complete the design and assist the contractor with installation details.

This project was in a plant site in Illinois during the winter and early spring. It snowed the day I arrived, and it snowed the day I left with plenty of winter weather in between. I learned that concrete could be poured in very cold weather if the area was heated and the concrete was prepared properly with the correct add mixtures. This was not a project resulting from a fire or explosion but was on a short schedule and required a fast turnaround on all decisions and designs. This was another example of observing the design produced and being installed in the field, which highlighted the problems needing correction and the good design practices that should be followed.

I was also called to the field for a project that experienced wind damage during the installation of a pre-engineered structure. Several of the main rigid frames and the connecting structural members had been twisted and damaged beyond repair. At first observation, it was obvious that the damage had occurred because the contractor had installed the rigid frames without the proper bracing and had gotten ahead of the lateral supports, which were needed to properly brace the frames. Overnight, the high winds in the area had damaged several frames and had to be taken down and replaced with new frames at the expense of the contractor who had not followed the proper procedure for the installation of the pre-engineered structure.

Trips to the field continued to present new problems that were educational but often similar to previous problems. Some issues were solved readily due to past experiences and the knowledge I was gaining by each new trip. This has continued throughout my career, and I am always seeing similar problems and learning from each one of them.

CHAPTER 13

Promoted to First Line of Supervision

My company provided engineers for the various plant sites in the maintenance, operations, and management groups as well as the various R and D and pilot plants. This provided an opportunity for transfers within the company to gain experience and knowledge of the full range of company activities. One of the more common areas in which engineers were transferred into was the construction group. This could be a permanent move, or it could be for short periods of time for the installation of a new or expanded plant facility.

My current supervisor had been selected to transfer to a plant site for the construction coordination of the capital projects at the site for an extended period of time. After I had been in the design group for several years, having experienced short field trips to solve problems and gained experience in a large variety of designs, I was promoted to the first line of supervision and was placed in charge of a small group of engineers and designers.

This was a new experience for me quite unlike the previous work I had done. I had always been responsible for my work but was now responsible for the engineers and designers reporting to me.

ENGINEERING

Since it was a small group, I continued to do my share of design. The task of letting go and expecting someone else to do the work was not easy and is one of the hardest things for a new supervisor to accept. The question is asked, "Why should I let someone else do the work when I can do it faster and better than they can? After all, I will be held accountable for the work."

Most new supervisors try to do the work themselves but cannot keep up without the help of the team working for you. It is the job of the supervisor to guide, teach, and supervise those working for him. Some first-time supervisors catch on very fast and learn to delegate the work and supervise others to succeed as a group. Others try to keep doing all the detail work and never let their group members learn to do the tasks themselves. Most successful supervisors learn to delegate and succeed as a group. I eventually learned to delegate, but I wanted to hold onto the work and do some of it myself. As the group grew in size, I realized that the best way to succeed was to let those working for me and with me to succeed, allowing me to enjoy the success of the group.

When you have a group of younger, less-experienced engineers and designers, it is important to use the skills they have, combine them with others who have different skills, and fill in the gaps with your experience if needed. This will also help the younger employees to gain additional experience and be cross-trained by working with those of different skills. Most of the supervisory skills I obtained were through observation, determination, and sometimes, desperation. I made some mistakes, but I also had a supervisor who guided me in the correct way to function and let me learn by doing.

Engineers and designers were placed on different projects so they could develop different skills. I would place an engineer or designer on a large project but task them with one specific type of work such as pipe support design. The engineering and design tended to be similar with each support but was repeated over and over so the employee would practice the same skill and gain a sense of familiarity. I could check their work on one support and offer suggestions and guidance, but they were doing the work and learning. Then I would switch up and let them work on the foundations with repetitive practice. If the project was large enough, they could be switched to another task with multiple engineering and design opportunities.

Another skill-learning practice was to assign the employees to a small job that required multiple skills and let them do the entire job. This allowed them to visualize how all the engineering and design functions were related to each other and also to gain a closer insight to the client requirements. On small projects, the engineers and designers were asked to go to the field to gather information, discuss issues with plant personnel, and complete an entire design package. Since these were smaller jobs, it was easy to assign one person for the engineering or designer functions so they could visualize and complete the entire project from start to finish.

The next natural progression was to assign engineers and designers to large sections of a plant design such as a process area, a tank farm area, etc. that had tasks requiring multiple skills. Finally, engineers and designers were assigned as leads for large projects who had other engineers and designers working for them. All these assignments required the oversight and guidance of the supervisor but allowed the employees to progress at their own speed and capability. When employees had gained sufficient experience and proven their capabilities, they were sent on short field assignments and sometimes assigned to the field office for a project in progress. At this point, the supervisor made contact and provided guidance as requested but essentially the employee functioned with little or no supervision. Employees at this point were usually on their own reporting to a job site or a project team but still did not have the supervision of others under them.

Supervision of a design group involved a lot more than just the engineering and the design functions. Once a project was under development, it required an estimate of manpower and materials, a schedule, the assignment of appropriate personnel, and a full understanding of the project scope and design group requirements. The project scope determined if additional third-party resources were required or if the design group could handle all the tasks. Items that were often needed were geotechnical reports, large-scale site surveys, and the use of turnkey vendors for specialized equipment and functions.

If the project was a plant extension or renovation, existing geotechnical reports and site surveys could be available and were used rather than requiring new work. If the existing reports and surveys were not available or were not adequate, a specification was required to issue along with a request for bids to secure the information prior to detailed design. Turnkey equipment or process packages were handled like any other equipment, and foundations and supports were provided as required. The designs for these items were a part of the schedule including the specifications, the request for bids, the approval period, and the receipt of approved drawings from the vendor to be used for detailed design. It was the responsibility of the design supervisor to handle these tasks and to guide the engineers and designers reporting to the supervisor.

Once the scope of the work was known, an estimate of the required manpower was provided. I used a combination of methods to arrive at an estimated number of engineering and designer hours required for the job. A determination was made of the number of drawings that would be required to show all the civil/structural/architectural work for the project, and the number of hours required for each drawing was estimated. Simple drawings were allotted a smaller number of hours, and complicated drawings received more hours. The number of hours were based upon past experience. This method provided a total number of engineering and designer hours for the drawings.

To that number an estimated number of hours to develop the project, the number of hours required for review and checking, and the number of hours required for issuing the drawing packages were included to arrive at a total number of manhours for the project. The second method is based on the type of project and the total size. A determination was made for the engineering cost as a percentage of the overall project cost, and this number was multiplied by the historical percentage of the civil/structural project costs, adjusted as needed based on the complexity of the project. The civil/structural cost was divided by the average cost per manhour to obtain the number of manhours.

The third method looks at the total design time for the civil/structural design group and the number of personnel available for the project. This could require adding manpower, working overtime to complete, or only using the manpower assigned part-time. If the manpower assigned and the schedule time available for the project are compatible, then the number of personnel assigned multiplied by the number of weeks in the schedule will provide another estimate of the required manpower. The manpower will tend to fill the time available.

If there is more manpower available than required to do the project, more time will be spent on alternate solutions and rework. If there is less manpower than required to do the work, it will tend to be conservative and cost more because there is no time to second-guess or look for alternate solutions. Having arrived at these different methods to estimate the manpower, an appropriate estimate can be determined comparing the estimate numbers and selecting one that is in line with one of the other methods.

There are more methods used, such as the number of manhours required for a piece of equipment or the manhours used on a previous project that was similar and adjusting the number for any variations. The important thing is to arrive at an estimate of the number of manhours required in which you have confidence and you have backup for your estimate. The supervisor will have more experience and knowledge and will be able to guide the younger engineers and designers in the preparation of engineering estimates for the group.

The importance of a good schedule cannot be overemphasized. The engineering department schedule for a project may be a combination of inputs from the various groups, or it may be provided by a competent scheduler in the department. In either case, the schedule must be reviewed and accepted by the individual design groups. If the project is large enough to justify, the schedule is usually provided by the department scheduler and utilizes software for project management and scheduling.

ENGINEERING

For a proper schedule to have real value, it should include all groups with input to the schedule. This will include all design, construction, procurement, cost control, maintenance, operations and management groups, as appropriate. There are links between all groups, and this must be considered in the overall schedule. The civil/structural group is usually concerned with the information required for the job, such as geotechnical reports, site surveys, equipment information, and preliminary information from the other design groups requiring supports or coordination.

After allowing time for design, there must be points in the schedule that allow for review with other disciplines and the client. There must be time allowed for checking and issuing of the design packages. If the information required is not received when needed, there will be an increase in the overall schedule, and there will be float time waiting for the information. If some other group requires information before it can be produced, that activity is on the "critical path" and may be sped up by adding more manpower but may not be able to be accomplished. The schedule will, therefore, reflect the increased time.

Usually, the schedule is reviewed and modified for several rounds in order to get the best schedule possible. This is a living document, and whenever something does not occur as planned, it is noted as a "schedule slip." Then the schedule is modified to reflect the change. To keep the final completion date, the schedule may be shortened to counteract the slippages by reducing any activity that is on the critical path, if possible. It is the responsibility of the design supervisor to review, approve, monitor, and maintain the portion of the schedule that covers the civil/structural activities.

Material quantities may be provided as a rough estimate at the beginning of a project but cannot be accurately estimated until the preliminary design has been done. Once the preliminary estimates of cut, fill, concrete, steel, etc. have been made, they are tracked and updated on a regular basis so the overall project estimate for material quantities are validated. These quantities are provided for a material takeoff estimate and will be produced as needed for "order of magnitude estimates," "plus or minus appropriation estimates," "definitive estimates," and "scope check estimates." Each estimate requires a confidence factor for the values reported and becomes more stringent as the design is developed. The supervisor must ensure that the material takeoff quantities are updated and reported accurately.

The assignment of personnel is extremely important to a project. If it is a small job, the supervisor may assume the functions of a lead engineer and guide the less-experienced engineers and designers assigned on the project. For larger projects, a lead engineer will be assigned. It will be the lead engineer's responsibility to review the manpower estimate, the schedule, the material estimate, and the manpower assigned to the project along with the supervisor. The lead engineer (or lead designer) will act as the supervisor's subordinate and perform the daily activities of the supervisor for the particular project and will review the activities with the supervisor on a regular basis as required. In addition to the lead, there has to be engineers and designers assigned who have the capability to perform the work on the project with or without guidance.

If there is limited knowledge or experience for a particular activity, the lead or the supervisor will step in and guide the personnel in the particular task. Depending on the schedule, there will need to be varying amounts of manpower at stages of the design, and the supervisor should adjust the assigned personnel to reflect this. A properly executed project that has no problems and everything occurs as scheduled is very rare, and the supervisor should be prepared to make modifications and adjustments as needed to keep the project going smoothly. An experienced supervisor will look ahead and anticipate problems and be prepared with solutions before the problems become evident. There is an old saying that goes "If nobody says anything, you must have done it right." If you do not do it correctly, you will hear about it.

A group supervisor should anticipate certain activities. If there is a structure or equipment that requires a foundation, a geotechnical report (new or existing) will be required. If the plant is expanded or developed in a previously undeveloped part of the site, a site survey (new or existing) may be required. If a large API tank is to be added, it will need to be field erected and may change the order of construction and, therefore, the order of design. If there is a large structure to be erected, it may alter the sequence of construction and, therefore, the sequence of design issue due to crane placement and accessibility.

If piles are to be used to support all major loads on the project, the entire pile construction package may need to be issued at the same time due to the cost of mobilization/demobilization of the crane and the accessibility of the crane. If the time required to fabricate the structural steel for the project is excessive, the structural steel package may have to be released early in order to meet the construction schedule. If the control room/office building is provided as a design-build package and bid out, the preliminary drawings for the building and the specification for the building will need to be released early in order to allow time for the successful bidder to design and install in the field.

If the project requires a large warehouse space to contain equipment for the project until it is installed and the project scope provides a large warehouse for storage of the product, it may need to be issued at an early date to be completed in time for use as a temporary storage for the incoming project equipment. This is only a small portion of the insight and possible modifications that may occur on a project that the design supervisor must anticipate and react to in order to keep the project running smoothly. This is where the experience of the design supervisor is vitally important.

The reason I know about these possible adjustments is because I have been challenged with these and more in the role of a design supervisor. Every time I solved one of these problems, they were filed away as an experience to be relied upon time after time. After seeing the same problem multiple times, I was able to recognize it and have a solution ready even before it became obvious that it was a problem.

In order to maintain control of the design effort, weekly project meetings are usually held with representation of all participating design groups, estimating and cost control, construction, and management and may involve the plant maintenance, operations, or client representatives. There is an agenda published that will consist of reports from each group participating covering the work performed during the last reporting period, the work planned for the next several weeks, a list of information required (including the date requested and the person responsible for addressing the request), and any other items affecting the scope or progress of the project.

Problems are identified in the weekly meetings and assigned to personnel to review and provide solutions. These problems are not generally solved in the meetings because of time restraints and research involved to find a workable solution. The use of 3D models or 2D drawings may be used to present information to the meeting attendees but is limited due to time restraints. Meetings are run by the project manager. Copies of all reports, concerns, requests, etc. are published and shared with the full project team members. Project management may request each participating group to provide reports prior to the meeting to be shared with the meeting members and keep the meeting short. A typical meeting would last for no more than one hour unless there was extra information to be shared.

First-line supervisors were expected to handle all problems that fell into their general area or classification. Process engineers were concerned with problems of the plant process, materials of construction, and the material balance, making sure the facility could produce the proper outcome if constructed and operated in accordance with the drawings and specifications. The electrical engineers handled the power, grounding, and lighting for the plant. The instrument engineers provided the proper controls to monitor the plant and ensure the plant operated in accordance with the requirement at each step in the process. The mechanical engineers guided the mechanical and rotating equipment through fabrication, installation, and start-up for proper operation.

The piping engineers and designers provided the piping systems to move the process flow through the plant and ensure that the flow rates and materials of construction were properly applied. The job of the civil/structural engineer was to provide the necessary foundations, structures, and supports for the equipment, piping, electrical, and instrumentation required to operate the plant. Our group also provided any architectural buildings or enclosures for the operation of the plant. Whenever a need arose that was outside the areas covered by the process, piping, mechanical, electrical, or instrumentation groups, it was generally the responsibility of the civil/structural group.

Our group handled the large majority of the fire marshal issues since they involved the buildings' and enclosures; exit and egress requirements. We usually took the lead, along with the piping group, on most site inspections since they involved structures and foundations but also involved equipment and piping. The equipment and piping inspections were assisted by someone from that particular discipline or turned over to them for evaluation. We conducted general material takeoffs for things that did not fit into any category.

Most of the other disciplines had set categories of work descriptions such as equipment list, instrument list, piping line list, piping and instrumentation diagrams (P&IDs), process flow diagrams (PFDs), electrical equipment list, electrical one-line diagrams, etc. Most of the civil/structural categories were all built from scratch. The general arrangements were provided as a start. But all the structural steel, the concrete foundations, drainage, catch basins, roads, railroads, dike walls, concrete paving, and architectural walls and roofs, etc. were all developed and prepared as a unique design for a particular plant. This made it much easier to develop and describe a design outside of a normal category for the civil/structural group.

We approached the problem in a similar way by using structural members such as steel, concrete, sections of drain pipe, etc. and joined them together to form a complete design for the solution of a particular problem. This was another reason for the first-line supervisor of the civil/structural group to be knowledgeable in a wide range of skills and experiences. It was a challenging position but one I quickly learned and shared with those engineers and designers who supported me in the group effort.

Some projects were executed smoothly, but others presented challenges and issues that required quick responses in order to keep the project moving along. As the supervisor, I enjoyed the success of those working with me. But I also had to take responsibility for those outcomes that were not as successful as planned. When our design effort was weak or failed, a timely response was required to get back on track and keep the project moving. Sometimes the issues were entirely out of our control, such as receiving late vendor information or a change in vendor data that required a change in the design. Other times the issues were totally in our control but we had come up short.

A design we had proposed early on that had been accepted by the project team was occasionally found to be inadequate in the final design and required a major design rework, causing a delay in the schedule or causing other disciplines to redesign their work packages. This was never a good thing, but it had to be accepted and corrected and the project brought back in line with the budget and schedule if possible.

Lessons were learned by the group members and especially the supervisor who, by experience and knowledge, should have seen the possible error and corrected it prior to the final design. I tried not to blame the individual who made the mistake but viewed it as a mistake that I had let slip through, and we all worked as a team to correct the problem, resolve to never let it happen again, and move on with increased knowledge, experience, and determination to always do a better job the next time.

CHAPTER 14

Additional Duties as CADD Manager

The engineering department began looking at new technology to produce designs and drawings in order to keep up with our competitors. We had been using software packages to analyze our structural designs since I began work. At first, they were time-share batch processes that were sent off for analysis, and we received output prints from a FedEx delivery service. This later on evolved into stand-alone software packages that could be run on department computers.

For the first dozen years or so, we continued to produce drawings on drafting boards with paper and drafting lead. In the mid-1980s, the department started a search for a computer-aided design and drafting (CADD) system that would produce construction drawings electronically to enhance our speed, accuracy, and professionalism. At the time, it was decided that a part-time position of CADD manager was needed to direct the efforts of the new system and manage the CADD operators as they learned the new system and began to produce drawings to replace the board drawings. The new part-time position was to be filled with a person who either knew a lot about computers or who knew a lot about engineering.

GERALD W. MAYES, PE, RETIRED

There were not a lot of people interested, but I always wanted a challenge. So I offered my services as the CADD manager. I was currently the group leader of the civil structural design group with approximately six to eight engineers and designers working for me in the group. My qualifications for the position included a dozen years as a civil/structural engineer with field experience and a limited knowledge of computers. My thesis was based on a computer solution for circular tank foundations, and I had used computers about as much as anyone else in the department who had any experience in engineering design. I was not proficient in either area, but I was also not totally unexperienced in either area. I think the reason I was selected for the position was that no one else wanted it. Either way I accepted the part-time position and began to assist in the development of the system.

The first task was to select a system that provided the department with the best value that would meet our needs. We considered mainly three systems: AutoCAD, MicroStation, and Computervision. MicroStation was a well-developed large system mostly used by highway departments and other large entities but did not provide a package that was suitable for the chemical industry and our specific needs. AutoCAD was a newcomer that was suitable for smaller projects and designs but was not fully developed for 3D designs at the time we reviewed them. Computer Vision was a company out of the Boston area that had all the capabilities we needed, including multiple 3D discipline design and interference checking. They were not as well-known but had greater potential for our needs in the future. Computer Vision was developing a Windows-based system that was beyond the other competitors at the time they were reviewed. The search committee consisted of several members of the department management team and myself.

Computervision was selected as the vendor for the new CADD system, and we began to prepare and train for the technology. I was sent to Boston for training classes on the new system. I had two weeks on the basic architectural and engineering course (AEC) for operation of the system. I was also sent for a one-week course for the system management of the system in order to maintain the system and install updates as they became available.

The first tasks to be produced were typical details and piping and instrumentation diagrams (P&IDs). The typical details were produced based on the basic AEC courses, but the P&IDs required another training course. P&IDs utilized the built-in equipment symbols provided by the system and connected them on the drawing using intelligent lines from one symbol to the next. Each symbol had related intelligence such as name, number, size, etc. The next type of drawing to be produced was a piping drawing. The piping drawings required valves, pumps, instruments, flanges, etc. that were provided as a generic model in the database. To include a valve in the piping drawing, the appropriate type of generic valve was selected from the database and modified for material of construction, diameter, type of flange, etc. These modifications were used to appropriately size the valve and locate it in the piping system. The intelligence provided for this valve could later be used to provide an electronic material takeoff of valve quantities. This type of modification and intelligence was provided for each item on the drawing including piping, valves, instruments, flanges, insulation, supports, gaskets, etc. If prepared properly, the system was very powerful and would produce great results.

Just like any other major change, there is resistance and a learning curve that has to be expected in order to achieve the desired results. The first major obstacle encountered was the delivery of an older system that we had received due to delay of the anticipated system. We had purchased a Windows-based system but received a temporary system using a light pen to select the objects from the database rather than using a mouse and selecting from the screen. It also included a computervision operating system (CVOS). This took some adjusting but was workable, and we started producing typical instrumentation details on electronic CADD drawings at one work station.

I had a separate manager station that I used for maintenance and updates that showed text on command lines but had no graphics. Eventually, we did receive the new work stations with icons on the work station monitors and a Windows-based system. The new system was based on a Unix operating system using C language. This was different from anything I had seen before, and I was once again required to go for additional training on the new system and language. We now had new work stations, a new operating system, a new programming language, additional CADD operators, and high expectations from management in the implementation of our new CADD system.

I quickly learned the basic programming and commands of the CADD system were based on the English grammar system of nouns, verbs, adverbs, adjectives, prepositions, conjunctions, etc. If you wanted to draw a line from point A to point B, which was x distance apart, you would type a command to draw a straight line from start point A to ending point B and enter a Return to complete the command. You could also specify a starting point and a distance of x in the direction you wanted the line to appear.

With the development of Windows with icons and menus on the screen, you now could simply select the correct items on the screen drop-down menus for Draw and Line and then click on the correct starting and ending points. In the above example, the noun is represented by Line, the verb by Draw, the adjective by Straight, prepositions by From and To, nouns by Point, adjectives by Starting and Ending, and conjunctions by And. All the command structure is developed in the same manner.

If we needed to modify the drawing database, we might have chosen to use the command modify from the menu. This allowed us a choice of commands such as Erase, Copy, Move, Mirror, Stretch, Array, etc. by selecting an object, the type of modification desired, and the specific modifiers needed to complete the modification. There were additional menus used to insert figures or blocks in the drawings. This allowed us to provide specific notes, weld symbols, dimensions, and other items to complete the drawing.

The use of three-dimensional drawings with intelligent databases further enhanced the use of the CADD system. If drawings were constructed accurately per accepted procedures with defined control points, such as elevations and dimensional distances to equipment and structure column centerlines, the 3D model could be relied upon for interference checks and verification for the accurate placement of the items shown on the drawings to be constructed in the field. It was very important for the CADD operator to connect all items to established known points in space and use tolerances accepted by the project team for accuracy.

Computervision had promised we could be producing intelligent drawings in a few months, and that was the expectation of management. Due to a lack of communication, the intelligent drawings promised by Computervision were based on their database intelligence already programmed into the system. However, the department management's request was for intelligent drawings that were customized to the department's view of previous expectations. This could be accomplished but took a lot of time to modify all the databases, and this was not acceptable to management. Neither side was wrong, and it was certainly possible. But the time required was not adequately communicated.

The vendor sent open-reeled tapes for updating the system every few months, and the system had to be shut down for at least a day in order to replace the old files with the new files. Some of the tapes would run for an hour or more to complete their tasks, and we typically received a half dozen or so each time. There was no internet or phone line that could be used. It had to be updated manually each time. There were usually some small glitches each time we received an update, and we would not know if it was a clean update until the new system had been run for a few days.

There were occasions when the electronic drawings were damaged or lost and backups were required. There was a system set up for backups that required multiple backups of the drawings produced or for those in production so the minimal amount of effort would be required to recover the drawing(s). There was a daily backup at the end of the day that copied everything that had been saved on the system since the last backup the previous week.

These daily backups were maintained for Monday through Friday and only overwritten when a full week of daily backups were available. At the end of each week, there was a weekly backup that backed up everything since the last monthly backup and was only overwritten when a full month of weekly backups were maintained for each week of the month. At the end of each month, there was a monthly backup that backed up everything since the last yearly backup and was only overwritten when a full year of monthly backups were maintained for each month of the year.

At the end of the year, a full backup was made of all data that was produced during the calendar year, and it was kept and was not overwritten. Because of the large amount of data stored on some of the backup tapes, the system required multiple tapes copies, which were labeled as multiple yearly, monthly, weekly, or daily. This allowed the system to be able to recover all data with no more than a few hours of production loss for any occurrence. To recover a drawing that was started several months previously, even in the previous year, the backups from the previous year, previous month, previous week, and the previous day are used to recover the data. Only the information from the current day, which had not been backed up yet, would need to be input to fully recover the drawing.

In addition to the regular tape backups, there was time spent on the system trying to correct problems the CADD operators were having. Some were simple to correct, and others proved more difficult. There was a lot of temporary files that were created during the production of the drawings that needed to be cleaned out and removed on a regular basis. The complete system utilized a pen plotter for producing the drawings when it initially began and later on utilized a dot matrix plotter and, finally, a laser plotter. The initial pen plotters worked well when they were working but produced a lot of headaches when they were not working.

The plotters used liquid-filled ink pens to transfer the electronic data from the computer to the drawing and used different pens for different line thicknesses. All drawings were initially printed in black ink. The computer chose the pen thickness it needed and began drawing lines, circles, arcs, etc. in a haphazard pattern, known only to the software programmer, and would switch pens as needed to complete the drawing. If the ink pens ran out of ink or became plugged, it was often during the plotting of a drawing. The pen then had to

ENGINEERING

be cleaned or replaced and the drawing replotted. Maintenance of the system including the work stations, the manager station, and the plotter was my responsibility and required several hours each week to perform. I shared my time between the CADD system and my other duties as the group leader of the civil/structural group.

One of the best attributes of the CADD system is the ability to check for interferences. When I first started in engineering, there was usually a small-scale physical model constructed for major projects in order to visualize the plant and the interaction between different disciplines, but it could not always detect interferences. With the CADD system properly utilizing 3D computer models and with all disciplines participating, the interferences could be found and corrected before the plant was built.

The first few times we ran a complete interference check on a major plant model, it would take all night to complete, but it would accurately detect and identify the interference. Not only would it indicate hard interferences such as a pipe going through a structural steel member, but it would identify soft interferences such as a valve handle that interfered with a structural member or piece of equipment in its present location. The valve handle could be rotated to a new position, and the interference was eliminated. The accuracy of the interference was due to the accuracy of the data input into the CADD model.

If details were to be repeated in prior engineering design drawings, it could either be referenced or redrawn. The use of a CADD system allowed the designer to copy details and repeat them at whatever location was needed. It also allowed the designer the capability to reproduce a portion of a drawing at a larger scale without having to redraw the detail. Similar details could be repeated on a drawing much faster and more accurately such as repeating the foundation outline of a pipe support on a location drawing. Anything that could be drawn could be copied, mirrored, stretched, or manipulated for the purpose of providing a fast and accurate drawing for review or construction. With the advancements in engineering and design today using CADD systems, most of these abilities are taken for granted, but they were not always available for use until the technology was developed and its widespread use implemented.

There were new problems that resulted from the use of CADD systems that had not been present before. When copying a specific note from one location on a drawing to another location and not changing the text, if required, it would result in an erroneous note unless it was checked and corrected. If view ports were improperly cut, there might be a portion of the view that would not be visible through the view port. If the scale of a view is changed to include additional information and the text is not corrected on the 2D drawing, the scale of the view will be incorrect. If any lines, details, text, or dimensions are placed on layers that are turned off, they will not appear on the printed copy of the drawing. There are numerous other problems that can be the result of using an electronic database for creation of construction drawings, but all can be eliminated if the drawings are properly prepared according to established procedures and are checked and corrected as needed.

By using a developed database of standard details, notes, and schedules, a specific tailored drawing can be produced using details that have been engineered and checked. These drawings can be produced quickly and accurately and will improve the efficiency and effectiveness of the construction package. We began to include these type drawings in each set of construction drawings and were able to reduce the time and effort required to produce a complete design package.

When the department produced drawings solely on the drafting boards, blank sheets with the company logo, title block, sign out blocks, and borders were produced by a local printing shop for use as a starter sheet for any drawings produced by the engineering department. Now we had the capability to provide an electronic copy of our standard sheets. I copied every line symbol and text from the standard sheet and placed it on an electronic drawing, which was issued to the printing office. From that point on, all drawings produced in the department were placed on the electronic standard sheet for CADD drawings and printed copies of the standard CADD drawing for use by any disciplines that had not transferred totally

ENGINEERING

to the CADD system for all drawings. Within a few years, all drawings were produced on the CADD system and the drafting boards were used to review drawings and, occasionally, to revise an old board drawing provided by a client.

This entire transformation of drawing production occurred during the late 1980s to the early 1990s. Eventually, my company converted all the Computervision files over to AutoCAD files. AutoCAD had become the vendor of choice by most companies due to its cost, convenience, and widespread use in the industry. AutoCAD now had good reliable 3D systems and the reasons for passing over them at the initial selection had been removed. In my own company, most of the satellite plant and remote facilities were buying AutoCAD systems, and to be compatible, we made the change to AutoCAD.

CHAPTER 15

Transferred to Construction—the Good, the Bad, and the Ugly

I had always enjoyed my time spent in the field reviewing and checking current design projects as well as working on special projects that covered several weeks to several months. In the fall of 1989, I was given a chance to transfer into the construction group and supervise projects in the field on a more permanent basis. I was transferred to a chemical plant in South Carolina and was assigned as a construction coordinator for all capital projects at the facility. This assignment carried a lot more duties and responsibilities than I had previously performed. The assignment became available when another engineer scheduled for the assignment declined the position. I was transferred within a few weeks of the offer and began my work while staying at a Holiday Inn near the plant.

Each plant site within the company had a small internal engineering group, and I was assigned to the group at the site reporting to the engineering group supervisor with an additional reporting responsibility to the project manager for the plant site who was located in the main office in Baton Rouge. The local plant projects were supervised by the plant engineering group, and the major

ENGINEERING

projects were supervised by the home office project manager. Even though I reported directly to the local engineering supervisor, some of my major projects were reported directly to the project manager in the home office. I had to keep both my supervisor and manager up to date on progress and problems occurring in the field.

When I arrived at the site, I was introduced to two contract construction coordinators who reported to me and were assigned various projects. A third contract coordinator who was assigned to me was not on-site at my arrival, and I met him at a later date. The third coordinator left after a few weeks and moved to another company, so I did not work with him to any extent. The current jobs were divided between the three of us, and we all supervised the current construction projects with the overall direction and accountability being my responsibility.

When I arrived, there was some tension between the local plant engineering group and the home office engineering group. The local engineering group had recently been placed under the home office and, in order to equalize pay and benefits, adjustments were made. Most of the employees affected were provided with additional benefits such as vacation, sick leave, etc. but also had their pay adjusted to come in line with others in the company. As human nature would have it, most employees only looked at the pay adjustments and not the extra benefits they were receiving. The adjustments and benefits had been reviewed, and in most cases, the employees were better off afterward but did not perceive the improvements. In addition, the announcement of the adjustments for the employees was handled very poorly, and it created a very tense situation.

Since I was a transfer from the disliked home office, my arrival was looked at with suspicion and animosity. I was even told by one person privately that I would probably not even make it there a full year. I did the only thing I knew to do. I ignored the perceived friction between the groups and treated everyone as my best friend and helped them in any way that I could. My relief came after only a few months when the project manager of the plant site was changed and I was being asked how to relate to the new manager since I had knowledge of him and none of them knew anything about him. It was not very long until I was accepted by the group and made some very good friends as I worked with them.

One of my first challenges came within two weeks of arriving at the site. One of the contractors at the site was upgrading the controls for a building that contained a very lethal chemical. This particular chemical was used in the gas chamber for lethal executions and was tightly sealed in a concrete block structure with exhaust vents to disperse and diffuse the escape of any leaks into the atmosphere. The instrument/electrical room was separated from the main building by a concrete block wall, and the electricians were running conduit through the wall into holes that were either drilled or punched through the wall.

I was making rounds to check on the progress and asked the electricians how they were sealing the holes around the conduit. They replied they did not have anything to seal the holes. I could just see the headlines: *Multiple People Killed from the Escape of Lethal Gas at Local Plant, Investigators Looking at Newly Assigned Construction Coordinator for Responsibility*. Maybe that was a little melodramatic, but I was concerned.

It was late in the afternoon. I went to the plant stores, but they were closed for the day. I was able to find some silicone caulk but had no caulk gun and could not find one. I left the plant site and started down a two-lane paved road that I had never traveled on before in search of a caulk gun and came upon a country store about five miles away and managed to find two manual caulk guns that I purchased. There was no way I should have been able to travel down an unknown road and find a store open after dark that would have caulking guns for sale, but prayers do get answered. I went back to the plant and presented the electricians with two caulk guns and several tubes of silicone and asked them to fill all holes around the conduit openings before they left on their work shift. I felt like I had done my part in maintaining the safety of personnel once the building was placed back in operation.

The first couple of months I was busy trying to become familiar with the projects that were in progress and the new projects that were ready for construction in the field. I relied heavily on the two

ENGINEERING

contract construction coordinators, and they helped me ease into the job. Most of the initial work was for small plant projects. The local engineering group consisted of project engineers and discipline designers. The designers produced drawings mostly for piping and equipment changes along with instrumentation and electrical upgrades and included concrete foundations and structural steel pipe supports. The project engineers had varying backgrounds and provided any engineering design necessary for the small projects.

When a package was complete with drawings, specifications, and scopes of work, it was awarded to one of the contractors on-site by the project engineers who had a limit of $25,000 for issued purchase orders. If the project was more than $25,000, the project engineer issued additional purchase orders up to the limit as a revision to complete the work. Once the purchase order was awarded to a contractor, one of the construction coordinators was assigned to follow the work.

I noticed that when projects were assigned to a project engineer, they usually awarded the work to their contractor of choice and the construction coordinator of choice so there appeared to be several smaller groups within the local engineering group. One of the reasons I was selected to relocate to the site as a construction coordinator was to consolidate the construction group under one person, eliminate some of the duplicating effort, and help select a single site contractor who could perform all of the work in a more effective manner.

Some of the practices of the group were not effective for producing good design or construction packages, and I set about to change what I could. I discovered that certain designers would not work with certain project engineers, and this created a lot of inefficiency. This was not in my area of control, so I took what packages were issued to me for construction and followed the awarded contractor in the work. I did, however, have control over the assigned construction coordinators. They were assigned to the projects based on availability and capability.

It was the practice of the piping designers to route pipe, and often, they would flat turn the piping in the main pipe racks, which is not an accepted practice. Most experienced pipers know that pipe should be turned up or down when making ninety-degree turns in the main pipe racks and when exiting and entering the pipe racks to keep from blocking any other pipes that need the space to continue down the rack. I think it was probably easier to just turn flat when there was not much in the way, and since most of their work was local and small, it made no difference.

For a larger project to use the same rack, it made a lot of difference, and I encouraged them to develop a practice of not flat turning unless there was no other way. I also noticed that on many small piping or equipment location drawings, the starting and ending point was a nozzle on a tank, pump, or other piece of equipment and that there was no elevation given at either end. It was further complicated by notes on the drawing to field route the pipe from point to point. This resulted in a drawing that indicated piping from point A to point B that could not be located for future work unless a field trip was made to physically locate the pipe.

It was even worse when the equipment indicated on the drawing by an equipment number had been replaced with a newly numbered piece of equipment and the number shown on the drawing was located in the plant boneyard for removed equipment. The designers were very qualified in their respective areas, but they had a small-plant mentality that did not match with an organized central engineering department that was producing drawings that had to fit into the same pipe racks, tank farms, and process structures.

And for those of you who think I am only complaining about pipers (this is a common practice in the industry for pipers and civil engineers/designers to complain about each other), there were plenty of problems with the civil and structural packages also. A common practice when needing a tank foundation was to find a previous foundation and copy it. It did not matter if the tank was taller or had a different specific gravity for the contents, it only mattered if the tank diameter was the same. If the project engineer did not have a background in foundation design or did not consult with anyone, the drawings were issued and the foundation was constructed.

ENGINEERING

This was a very unsafe practice, and I made sure it was stopped. It was not my responsibility to correct design practices, but I was responsible for constructing the designs. Anything to improve the process and safety for the group was welcomed. Every major foundation was reviewed to ensure it had a proper design and that it was capable of supporting the tank and its contents. Another unsafe practice was to add onto existing structural supports and pipe racks without checking the support steel for capacity. Unfortunately, this is a common practice industry wide, and as long as unqualified maintenance personnel are allowed to make structural revisions in the field without qualified engineering oversight, there will be the potential for unexpected consequences. There were also common practices in the construction group that needed to be addressed in order to complete projects with quality and efficiency.

The plant contractors consisted of about five to six groups at any time and sometimes were more than six. There were at least three major contractors that could completely perform all the work on a given project, and there were other contractors who did specialty work such as pressure washing, electrical and instrumentation, painting, and insulation. The general practice of the local engineering group was to provide work for the on-site contractors so they could remain on-site and be available for work. This led to a lot of inefficiency, and the work was often awarded to the contractors before it was complete to keep them busy.

One of the electrical and instrumentation contractors had a core group of three people: a foreman and two journeymen. If enough work was available, it would keep three people busy, and if there was a lot of work, they would bring in a couple more people. When there was no work, the usual core people would come in with no pay and wait from seven a.m. until about ten a.m. to see if anything was available. If there was no work, they would leave for the day with no pay. To keep the workers available for possible work, the local engineering group would give them work ahead of schedule, which was inefficient. Whenever there was a large project (requiring a lot of workers) awarded to them and they could only find five or six people to work, the project completion was delayed.

GERALD W. MAYES, PE, RETIRED

One of the primary reasons for me to be assigned to the plant site was to consolidate the work under one contractor. This meant that some of the contractors who were very dependable and loyal were forced out of the plant because they could not handle the larger projects that were soon to come. I was part of a team that interviewed multiple contractors for the plant site. These contractors were expected to handle all the work presented to them and only bring in subcontractors as needed, such as a subcontractor to perform x-ray services and so on. The goal was to have one general contractor to manage all the capital construction at the site who could efficiently provide the required manpower, equipment, and resources.

After researching the prospective companies, it was reduced to three major contractors who were interested in providing the construction resources. Each contractor was asked for a proposal to include the organization of their company, resources available, safety records, proposed labor rates, and résumés of proposed personnel to be assigned. The reviewing committee compared all the proposals and selected a single contractor for the award. The work being handled by the various on-site contractors were phased out, and they were gradually removed from the site. A large portion of the various contractor trailers and sheds were removed or dismantled. Many of the well-qualified experienced construction personnel were hired by the new contractor and never left the site.

At this time, another problem was being solved involving surplus material held in the construction site sheds. It was a common practice at the site to spend any funds left on a project for material that could be used for a future project. The proper procedure would have been to return the funds to the project as underrun, but the site maintained their own slush fund storage. For the next project that needed material and was tight on money, the material was simply pulled from the storage and used as a part of the project. I was asked to eliminate the practice and the storage of materials.

Most of the material consisted of high-priced valves, fittings, piping, and controls that were common to the local plant projects. A salvage dealer was contacted, and the valves and fittings were sold to the dealer for about ten cents on the dollar. The plant received a check for over $10,000 for the salvaged materials (purchased value was over $100,000). Some of the other material such as pipe was traded to the plant stores in return for an air conditioner to be used on controlled-atmosphere enclosures containing sensitive material storage on upcoming projects. The current practice was to store sensitive electrical and electronic materials on a pallet covered with a tarp in the outside weather for several months until it was installed. After exposure to the elements, the equipment often did not work and had to be replaced or repaired prior to installation.

The existing construction area consisted of several open-air sheds used for storage, welding, and pipe fabrication of items for installation at the site. The sheds had not been maintained due to the lack of funding. I discovered that no temporary construction funds were added to any of the local projects to be used for maintenance or repairs in the construction compound. I requested funds to be added for temporary construction to each new project. All the major projects supported from the Baton Rouge office usually contained temporary construction funds.

After the other construction groups had left the site and cleared out their areas, I prepared to clean up and maintain the facilities that were left. Most of the sheds that were used for welding and pipe fabrication leaked every time it rained and had to be patched and sealed. Most of the metal roofing was intact and useable, but it required some screws, nails, roofing tar, and sealer to stop the water intrusion. This was relatively easy and was started as soon as any funds were available. One particular shed was enclosed. An air conditioner was obtained through trades with the plant stores group and was used to provide a place to store computers, electrical and electronic controls, and other sensitive materials until they were installed.

One of the major issues for the construction forces was a place to eat lunch. In accordance with plant rules, the construction workers could not prepare or keep food in the construction area but were allowed to bring in food in a lunch box only. Since there was no refrigeration, the foods were subject to spoil, and this created a safety hazard for the construction workers. Most of the construction workers ate at the plant lunchroom but were not allowed to eat until after the plant forces had eaten. The lunchroom was located at the opposite end of the plant site from the construction compound, and because of plant rules, the workers were not allowed to ride in the back of trucks to go to lunch but had to walk. The average time to walk to the lunchroom was approximately ten minutes with another ten-minute wait in line and ten minutes for the return trip. This meant that thirty minutes each day per construction worker was lost productivity.

At one time there were over two hundred workers. This amounted to a loss in productivity of one hundred hours per day worked or 5 percent loss for a ten-hour workday. I was allowed to handle a few projects as a project engineer, and I submitted a proposal to construct a facility in the construction compound that would serve as a lunchroom, meeting room, and restroom for the construction forces that would pay for itself in less than a year. It was approved, and a pole building with metal siding and insulation was constructed in approximately one month. The building had restroom facilities for men and women and was equipped with folding tables and stackable plastic chairs to seat in excess of 120 workers.

Lunch was handled in two shifts, and safety meetings were held with standing room only. But it was close and dry, and that was better than what the workers had previously used. I purchased two large commercial refrigerators for storing lunches and provided several microwaves for heating food. I brought in outside vendors to setup drink and snack machines at no cost to the plant or construction group. This facility provided space for general meetings and safety training for the construction crew. A shed was provided on one end of the building as a designated smoking area for those who smoked since it was not allowed in the plant site except at designated sites. All this construction area improvement was provided either by trading material that I had been directed to dispose of or by a properly executed project with no additional cost.

During this time, I worked very hard in separating any benefits I received from the job and the position that allowed me to have control over many people and contractors. I ate lunch with many contractors but always required that business was discussed and not

ENGINEERING

just the contractor trying to gain favor or obligation from me. I was once offered top-quality lumber from a demolition project that was the property of the contractor. I turned it down and bought lumber full of knots and imperfections that I needed from a local hardware store. I had stumps in my yard from trees that had blown down during a hurricane that needed to be removed. The foreman for the construction contractor offered to come on the weekend and dig up the stumps. I turned him down and found a backhoe contractor that was so bad and had questionable equipment that he would not be allowed on-site and hired him to do the work. It took him an extra day because of equipment failures, but no one could ever accuse me of trading favors for work.

I was told I would be getting a vehicle to move around the plant site and travel from home to work and return, but it never developed. I was provided with a gasoline four-wheel two-seat scooter for moving around the site, and it was more convenient. It had no windshield and no top, but it was cool in the summer and cold in the winter. It was especially cold in the winter when it was raining. Most outside construction was halted for rainouts, but a lot of work was being done inside buildings, process areas, etc. The rain did not stop my movements. I wore rubber boots and a slicker suit with the pants down over the boots and the top down over the pants to divert any water. My hard hat kept most of the water off my head, so I was ready to go, come rain or shine. The scooter would get me into most areas, but sometimes, I had to walk short distances to gain access to barricaded sites or restricted sites. If I was inside an area that experienced a flammable leak, I might have to leave the scooter and walk out and return later when it was safe to run the gasoline engine on the scooter.

My hours generally were from 7:00 a.m. until 5:30 p.m. with a half hour for lunch, Monday through Thursday for a forty-hour workweek. I usually arrived around 6:30 and left around 6:00 pm and was almost always at the plant on Friday. Sometimes the schedule required me to work on weekends or on nights. It was a plant requirement that when any contractor was on-site doing work, there would be a construction coordinator there for the duration. Many times, one of the contract construction coordinators would cover these instances, but sometimes I would be needed because of a project I was followings and so on.

I remember being at the site one Saturday morning at four a.m. inside a large tank inspecting a weld and wondering why I was there at such an odd hour. If there was work on a second shift, it was not unusual to be at the site at eleven p.m. It was not always that time-consuming, and there were times I took off for doctor visits, family affairs, etc. I was told by my project manager that I was to put forty hours per week on my time sheet, and it did not matter if I worked thirty hours or fifty hours. I kept up with my time at the plant site over a period of four years, and not including any time off, doctor visits, or other absences, I worked enough additional hours for a full year. This meant I worked an average of fifty hours per week for forty hours of pay, but it was understood and I enjoyed my time in construction at the site.

Typical construction projects were multidisciplined and involved expertise in concrete foundations, equipment setting, steel structures, piping systems, electrical power and grounding, and instrumentation for control of the process. Other minor projects might involve only piping and instrumentation or miscellaneous supports or platforms for existing equipment and processes.

When the construction package was received by our group and transmitted to the construction contractor on-site, the scope was reviewed and manpower was scheduled to perform the work. Each discipline of construction workers had to schedule their work and provide equipment, resources, and manpower to accomplish the overall effort required for the project. A construction coordinator was assigned to each project to follow the work and ensure that the proper scope was being followed and the work was done in a safe, efficient manner.

If the projects were small, a construction coordinator was assigned multiple projects to handle. In the same manner, the construction crews were working on multiple projects simultaneously. This made for an efficient work effort so that a crew could pour a small pump or equipment foundation on one project and move to another project while waiting for the concrete to set up so the forms could be removed. Piping crews might have to pull off a particular project while waiting for a piece of equipment or an instrument to be delivered on the site. An electrical crew that was running conduit outside in the pipe rack might have to pull off the particular task due to a rainstorm but might be able to work on the inside of a new switchgear building or control room.

Most of the construction projects were cost-plus, which meant that the company paid each worker for the hours they worked plus an agreed amount or percentage over the cost of the individual worker. The company also paid for the cost of any materials or equipment purchased by the contractor plus any agreed amount over the cost. It was the responsibility of the construction coordinator to monitor the work of the contractor and determine the hours and materials used on a particular project.

With the help of the construction coordinators working with me, the daily effort for each project was determined. Time sheets were signed each week that indicated on a daily basis the hours and project worked by each construction employee. Contractor purchases of material and small equipment required for each project over a set amount (I think it was either $500 or $1000) were required to have three bids, and I had to sign off on the requisition before the contractor could purchase. When the contractor submitted an invoice on a monthly basis for payment, it included copies of all the contractor requisitions and a copy of all the approved time sheets.

One of the very first things I did when I arrived at the site was to require the contractor to purchase time sheet forms on three-part paper. This allowed the form to be filled out and approved so I could tear off and keep the second copy in my files. The contractor had the original form and the first copy. The first copy was included with the monthly invoice, and all I had to do was compare the approved copy that was retained with the supporting copy attached to the invoice and the time of each employee was verified. Any charges for material and small purchases were verified in the same manner.

Occasionally, we would have a hard money project in which a contractor, usually a subcontractor, would be awarded a fixed price to complete a job and was paid no more unless the contractor could justify an increase due to additional or modified scope of the project. In this case, the construction coordinators were not concerned about the time or materials used for the project but rather the completion of the scope of work on time. Extra effort had to be made not to mix cost-plus and fixed-price projects in order for the contractor to not work on a fixed-price job and charge us for manpower on a cost-plus project. Regardless of the type or number of projects, it was our responsibility to keep track of the work being done by the construction contractor on-site and any subcontractors or specialty contractors required to do the work. We had to make sure the work was coordinated with the plant operation and maintenance forces for a smooth and safe project.

Most of the projects we constructed were similar in nature. A typical project consisted of the demolition of equipment, piping, controls, electrical service, and structures that was no longer needed or had to be replaced, resized, or upgraded in some manner. There might be some minor earthwork and grading required, especially if the project extended into areas that had not been utilized before. Foundations were installed; equipment was set; and piping, electrical, and controls were added to complete the process. Most of the new equipment, piping, and structures required either insulating or painting. In some cases, the structural members were shipped to the field with galvanizing applied in the shop and required no further surface coating.

Once the equipment and piping systems were installed, they had to be pressure tested. Some pieces of equipment were hydrotested in the fabricating shop, but the piping systems were typically hydrotested in the field after installation. The electrical systems had to be checked to verify the proper capacity and service. The control systems were checked in the field to verify proper operation, and chemicals were introduced to the piping systems to check flow rates and proper operation before start-up could begin. Each step in the process was checked and verified for proper operation and control. Once the start-up was complete with all problems addressed and corrected, the plant was ready for production and was turned over to operations. Construction forces were utilized during the start-up procedure for corrections and were pulled from the project once the plant was operating.

There were cases of nontypical projects that sometimes were expected and sometimes were a surprise. Anytime you are excavating soil from the ground in order to install a new foundation or drainage pipe, there is an element of the unknown. Occasionally, contaminated soil is discovered, and it has to be removed, replaced, or otherwise remediated. Underground pipe, drain lines, and electrical duct bank are often found and must be dealt with.

The existing water table may be higher than anticipated, and the foundations may require modification. Or the ground may need to be dewatered in order to install the foundation. Drilled and cast-in-place piles, auger cast piles, and driven piles may need to be relocated to miss underground obstructions. Soft spots in the existing soil not addressed in the geotechnical report may be discovered beneath a foundation that must be corrected before the foundation is installed. While installing a foundation adjacent to a previously installed foundation, it may be discovered that there is significant overpour from the existing foundation that must be removed to allow for the new foundation.

Not all problems are found below grade. Structural steel equipment supports may have to be modified to match the mounting holes on the equipment received when they do not match the support holes indicated on the certified drawings. When attaching new steel to existing steel, there may be existing conduit and piping that interferes with the connection of the steel. This can certainly be reduced by field checking the drawings before installation of the new steel, but it does not eliminate all problems, especially if the conduit or piping is field installed on the same project and there is not good communication between the design groups and the construction forces.

A large portion of controls and instrumentation sizing is done from cut sheets and standard drawings and is required early in the process for installation of the in-line controls in the piping systems. If any instrumentation is shipped to the field with a different dimension or if the piping system has a certain amount of field routing, there is always a chance of not fitting perfectly, and it will need to be adjusted in the field. This is an accepted practice and as long as it does not become excessive, it is an effective way of handling small corrections in the field. Minor corrections are usually made in the field, and larger issues are sent back to the design groups in the form of an RFI (request for information) for resolution.

I was given the opportunity to function as a project engineer for several smaller projects while at the site. In addition to the project to build a lunch and restroom facility for the construction forces, I was a project engineer for the completion of a project to remove asbestos from a plant area. The work was contracted out to a firm that specialized in asbestos removal, and I followed the work and made sure it was completed within schedule and budget. I worked on other minor projects as a project engineer.

One day I was called to the maintenance office for a meeting. I noticed I was the only person from the plant engineering group present since all the attendees were from the site maintenance or operations groups. A large glass-lined reactor had been damaged and had to be replaced. The purchasing group had located a replacement, but they needed to have it installed immediately. I was asked to locate a contractor for the installation on an emergency basis. The meeting was in the morning. I had a contractor there that afternoon to look at the area, and they began work that night. The plant maintenance group had already begun to remove the piping connected to the reactor and prepare it for removal. The area was so congested that the only way to remove the damaged reactor was to remove the supporting steel beneath the reactor and lower it down from the second floor and out of the building. The space was so close that the reactor was touching structural steel on both sides as it was removed from the building. Some of the insulation on one side was torn off when removing.

The outside contractor had brought in a crane for the removal and had provided enough personnel to make the swap out of the reactors. They had two crews that started early in the morning and finished late at night. The reactor was replaced and secured and supported by the second-floor steel. The piping and controls were reattached, and the entire project was completed with four hours to spare from the time we had been allotted to complete the task. I was called in because the plant management knew they did not have the capability to make the swap, and our construction organization had proved that it could move fast and preform under pressure.

While I was at the site, there were a lot of interesting projects. We installed a sewer upgrade in order to satisfy state EPA requirements complete with double containment, paving with water-stopped joints, and a new clarifier to treat the plant effluent. We constructed a training facility structure for the plant firefighters to develop and maintain their skills at fighting fires. We repaired a plant that suffered an explosion due to an equipment failure. We constructed a clean room for processing pharmaceutical grade chemicals as the main ingredient used in the pain relief field.

The plant maintenance, operations, and other blue-collar jobs were all unionized. The construction forces were all nonunion. This was not a problem, and the union and nonunion forces worked well with each other with no problems. While at the plant, a union contract was up for negotiation, and when no agreement was reached, the union forces went on strike. There were several construction projects at the time, and they were not affected unless there was need for a plant operator or maintenance person to perform a task. The operations and maintenance groups were being manned by supervision, and response to specific tasks were slow.

Before the construction forces could perform any task on an existing in-service piece of equipment, it had to be shut down, cleaned, and turned over to the construction forces. The construction forces were required to use a separate gate to enter the plant site, and work continued. For the management personnel, they were assigned to various duties, and folding cots were brought in to provide a place to sleep during the strike. Since I was over the construction forces and they were not on strike, I continued to go through the gate used by construction.

The plant opened up the cafeteria and started serving those people required to stay on the site. I was also able to eat there during my work shifts, and I did gain several pounds by eating multiple times a day. My first task after the strike was announced was to search the construction area for a bomb. The bomb threat had been called in but turned out to be false. After a few weeks, the contract issues were resolved and the work returned to normal.

I was able to expand my knowledge and experience of all the design disciplines by following the work in the field. I learned about double block and bleeds that were used to isolate equipment and instruments that might have to be taken out of service. I learned about pulling wiring through conduit, what could be done, and what should not be tried. I learned about the importance of setting and aligning pump and pump motors to prevent future problems due to misalignment. I watched field-erected steel members being installed and learned the proper way to connect and brace the structure during erection. I learned the proper way to hydrotest a piping system and where to install drains and high point bleeds to eliminate air in the pipe system before pressurizing.

I learned the proper way to position a crane for a heavy lift and allow clearance for removing and setting loads safely. I learned the correct way to land wiring on terminal strips. I had learned in college the proper way to calculate stresses in structural members. In the central design office, I put my knowledge to work learning to coordinate with other groups for a complete design package. The four years I spent in construction at the South Carolina site greatly increased my knowledge of how it all comes together. I was able to see how all the design disciplines did their part in providing a complete package, but most of all, I learned to appreciate each groups' contribution and understand how the sum of all the parts made for a complete engineering package.

During my stay at the plant site, another direct-hire employee was assigned to help me, and after four years, I was told that I was needed somewhere else. The employee who worked for me was promoted to the construction coordinator position. The company moved me and my family from South Carolina to a plant site in Illinois. We sold our home in South Carolina and purchased a home in St. Louis, Missouri, just across the Mississippi River from the plant site in Illinois.

We looked at homes both in St. Louis and across the Mississippi River in Illinois within driving distance of the plant site. The plant was near the river and was in sight of the Gateway Arch near downtown St. Louis. We decided to make an offer on the best home available and found one in South County, St. Louis, that was close to the interstate bridge south of the downtown area. It was a two-story four-bedroom home located in the suburbs that was great for me and my family.

I moved to the plant site a few weeks before my family was moved in July of 1993, and the first night there was during the high water of one of the worst floods in the St. Louis area. The news that afternoon showed the extent of the flooding, and I saw entire houses floating down the Missouri River and into the Mississippi River. Barges were sent adrift, and there were floating propane tanks full of propane drifting down the river. It was not the type of welcome I expected or enjoyed.

When I returned with my family and the moving van delivered our household goods, we started our life, and I continued my career in St. Louis. My wife and kids had time before school started to unpack and check out the area, and I reported to my new job at the plant.

The job was similar to the job I had left in South Carolina but different in many ways. This was a large project that received close supervision from my immediate supervisor who was located in Houston and the project manager who was located in Baton Rouge. I was the construction coordinator who was on-site every day but reported to my supervisor and project manager almost daily and in person in regular scheduled meetings that were held at the site as needed. In addition, there was a plant manager with whom I was constantly coordinating with every day.

I was the constant contact but directed by management on a continuing basis. There was an outside contractor responsible for the construction effort, and I was in daily contact with the outside contractor site manager. It was my responsibility to coordinate the construction effort with plant operations and maintenance forces for the construction effort, which was located in areas under full production. It was a full-time job coordinating all the different groups and I was allowed to hire an assistant to keep up with the paperwork, fling, copying, etc. My day was filled with reviewing the construction effort, coordinating with operations, maintenance, and the construction contractor. Even though I was heavily coordinated by management there was a need for someone on-site to handle the everyday concerns, and someone who could make a decision and keep the project moving. I had demonstrated at the previous site that I was capable and handling the construction effort and was transferred to this site to handle the single contractor working on one project. Just because I had some experience and had performed at the previous site did not mean that I did not make mistakes and I was corrected as necessary by the management supervision I received.

The external contractor awarded the construction for the project brought in a management team consisting of administration and clerical personnel as well as safety, warehousing, site engineering, and craft supervision personnel to manage the construction effort for the project. This was a union site, and all the skilled craft personnel were members of the local unions. This was a total switch from the previous site where all workers were nonunion. It was a mixed blessing. All the workers were highly skilled, but it took more personnel to perform the same task. I was not involved with the contract requirements or costs, and I was able to adjust to the union site with no problems.

The construction contractor brought in typical trailers and set up for the management and administrative type offices, complete with reproduction equipment, desks, chairs, tables, and filing cabinets for the offices and conference rooms. Trailers were used for the safety, site engineering, and support personnel. For the site fabrication facilities needed for pipe and minor steel fabrication areas, shipping containers were stacked two levels high in a U-shaped pattern, and a temporary roof spanned across the top of the U-shaped arrangement. To enclose the area and provide wind resistance for the welding operations, tarpaulins were hung across the open side.

ENGINEERING

The inside of the shipping containers was used to store parts and supplies needed for the construction effort. Temporary stairs were constructed to reach the shipping containers located on the top. An area was provided as a lay-down yard to store large sizes and lengths of materials such as pipe, structural steel, cable tray, conduit, wiring spools, and large pieces of equipment awaiting installation. The entire construction area was fenced and secured to prevent unauthorized access and removal of supplies and equipment. A trailer was provided nearby for the company construction coordinator with desks, tables, chairs, filing cabinets, drawing storage racks, and printing and copy equipment.

Regular meetings were scheduled with the contractor on a weekly basis and other meetings when required. Current progress and planned progress for the future was discussed, and problems were addressed. Usually once a month, there was a meeting held with attendance by the contractor key personnel, construction coordinator, company project management from Baton Rouge and Houston, and the project management from the plant site. In addition to the usual items being discussed, financial reports and projections were included. Due to lack of space for these meetings, they were often held off-site.

The site primarily involved the production of various oils, oil additives, and lubricants. This particular project involved the addition of equipment, piping, storage capacity, and supporting facilities for a new additives plant and encompassed most of the plant site in one way or another. There were storage tanks that were installed on new foundations, and this could be done early since it did not interfere with any of the production operations in the plant. The plant site was located near the Mississippi River and had a high water table. The site also had contaminated soil below grade due to the previous owner of the site's manufacture of chemicals for the government. Because of this, most foundations were designed to not remove soil from the site. These tank foundations were supported by driven square concrete piles. The first few months of the construction effort were involved with equipment foundations, dike walls, pump foundations, etc. that did not interfere with operations.

As the work progressed into the piping phase, there was a lot of demolition required in order to install the new piping. In each case, the piping lines had to be isolated, the product removed from the pipe, and the pipe had to be cleaned before removal. If the demolition could be isolated at a valve or flange, the separation point would have a blind flange installed and the service returned to the portion of the pipe that was left. Some demolition could not be done until the plant area was shut down and all operations were halted in order to make it safe.

Once a section of pipe was isolated, it was removed in sections by cutting it into manageable pieces with a portable bandsaw. If the line was safe, it could be cut into sections with a torch, but this was not the first choice for pipe that had contained oils, lubricants, and other flammable materials. A cutting torch could be used on waterlines, steam lines, and other piping that was deemed to be safe. For any demolition, there had to be a hot-work permit obtained from plant operations that specified what was to be removed and the method and tools used for the removal. The demolition was a slow process, especially in the middle of an operating plant, and required a lot of time in the project schedule to complete.

There was also a lot of demolition that involved electrical and instrument control systems, conduit, and wiring. Special care had to be observed in order to remove the proper equipment, controls, and wiring without affecting any of the systems that were required to continue plant operation. Demolition was required for concrete pedestals and foundations that were no longer needed due to equipment removal or relocation. Some demolition was required for existing dike walls that had to be removed or relocated due to addition tanks and equipment installation. Existing process drain lines were demoed to install larger diameter pipe. Existing pumps were removed, and the foundations were modified by extending the foundations, cutting the existing anchors off flush with the top of the foundation, and installing new post installed anchors for the new pump base.

Some demolition was needed to remove piping, conduit, and other equipment that had been abandoned or was no longer in use. This was a high priority since it did not require shutting down the

ENGINEERING

plant to remove. Demolition that could not be done while the plant was operating was scheduled to be completed during a plant shutdown. This usually involved removing a section of pipe and replacing it with a section that included a valve to isolate the piece in order to start the plant back up. After the plant was running, additional piping could be installed and attached to the piece installed during the shutdown without affecting plant operation.

If the piping tie-in was critical and could not wait for a plant shutdown, a "hot tap" was used. This was a method of attaching a section of pipe containing a valve, usually at ninety degrees, to another pipe by welding. After the welding connection is made, the opening between the pipes is drilled out, the drilled section is removed, and the valve is closed. Operations continue during the whole process, and now there is a connection with a valve that can be used to attach future pipe. Using hot taps is a common method of making tie-ins, but they are only used if necessary because of the high cost. The cost of the hot tap is weighed against the cost of shutting down the operations, and usually the hot tap cost is lower, especially if the shutdown is critical to the plant operation.

Most plant construction follows a logical pattern unless there are circumstances that require adjustments. Site work, foundations, underground piping, electrical conduit, electrical grounding, and duct banks are completed first, and any interferences are resolved. The next step is the installation of large pieces of equipment that rest on concrete foundations. Sometimes this is delayed due to equipment arrival at the site and can be installed later if it does not hold up any other construction. This is followed by the installation of the steel structures for buildings, open-air structures, and pipe racks. In conjunction with the structural steel, larger pieces of equipment may be placed on the steel supports as the structure is being erected for ease of installation. As soon as the pipe racks and the steel structures are erected and braced, the installation of major pipe lines, electrical power, and control wiring are placed in the structures and connections are made to the equipment and electrical and control houses.

Installation of motor control centers and control rooms are generally independent of the process structures but must be in place before the wiring and controls are attached. Any architectural siding roofing or protection is installed on the structures at the appropriate time to benefit installation and protection as needed. Minor pieces of equipment and controls are added once the structures and piping are in place to support them. In-line controls and equipment will be installed with the piping if available, and if not, they will be installed when received in the space left for them in the piping.

Once all the equipment and materials are installed, there remains a lot of "checkout" that has to be performed. Ideally, major equipment are pressure checked in the shop, but piping systems are usually pressured checked in the field. A section of the pipe is isolated between blind flanges. Low-point drains and high-point vents are installed in the pipe in order to eliminate all compressible air from the selected section. The section is pressurized in accordance with the project specifications, such as 1.5 times the maximum operating pressure, and tested to determine if the pressure holds for a specified time. If it does, the line is verified as being pressure tested and is recorded.

Pressure checks may be recorded on a copy of the approved PFD drawing to indicate what has been tested and what piping remains to be tested. Sometimes individual pieces of equipment are pressure tested in the field along with the piping if it is appropriate. This procedure continues until all the new equipment and piping has been individually checked and verified.

When everything has been installed, tested, and approved, the construction project moves into a new phase. This will require the personnel from the design office to visit the site to assist in the start-up. Product is introduced into the systems, and the piping and control systems are tested to make sure they function properly. All the electrical and control wiring must be tested to see if there is proper continuity and if the systems are properly installed and preforming as designed. The automatic valves and sensors are checked and evaluated. Adjustments are made as needed to ensure the proper flow rates, and times are correct for maximum operation of the plant. The control logic is verified, and the operators are trained in the proper procedures and sequence of the plant operation. Safety measures are tested and evaluated.

In a nutshell, the entire plant is given a test drive, minor adjustments are made, and the start-up is declared over. The plant then moves into full operational mode to produce the desired product. This is an especially cautious time since most accidents occur during start-up or shut down, and safety awareness is at its highest level. Once the plant accepts the new installation and agrees all its systems are working properly, the ownership of the project is passed to the operations group and the construction forces start their demobilization of the site.

There are always exceptions to the schedule and the planned activities during the construction phase of a project, but the closer the actual installation follows the agreed plan and schedule, the smoother the construction will go. Construction forces must always be ready to make adjustments and determine the effects of change on the manpower and schedule of the project and report them to upper management. If management chooses to wait until all design is completed and all deliveries are made on-site, most plants will not be built in time to serve the needs of the customer. There may not be enough space to handle all the equipment and material deliveries on-site. The effective project manager must be willing to accept small risks in the cost and adjusted schedule of a project in order to reap the benefits of a plant delivered to the client's satisfaction. It is the responsibility of the construction coordinator to be flexible and adjust to changes in the schedule and budget and complete the job safely and as efficiently as possible.

The construction coordination at this plant was different in many ways from the previous site. The construction site superintendent requested that he be able to provide pictures for regular reports made to his supervision. Since this was an outside company, the plant required that all pictures be reviewed and approved before releasing to the contractor. After a couple of weeks of having to review and approve the pictures, the company plant manager suggested that I get a rubber stamp with a blank for signature and date and place on the back all the pictures and review and approve them myself. This was easy and convenient to do. The only odd thing was no one told me what I was reviewing the pictures for and why I should reject them.

I used a little bit of common sense and satisfied the plant managers desire to remain in control of items leaving the plant and the contractors need for pictures of the construction progress. This was a union labor site, and I had to approach the contractors in a different manner. But the work was essentially the same. This site had to be developed as a construction facility complete with storage laydown yards and lockable storage trailers/fabrication areas as well as construction offices. This effort involved the construction effort for one large plant expansion while the previous site handles multiple construction efforts on a smaller scale. This site was my first time to coordinate a complete outside contractor brought in to complete all the construction effort. Like most construction jobs, the work came to an end, and the people working here went on to other projects, including myself.

CHAPTER 16

Back to Design

When the construction coordination job in Illinois was scheduled to close, I was initially told I could be assigned as a site design supervisor. While I was there, the company split into two separate companies, and the company that I reported to did not have any personnel located at the site. As the construction project was winding down, I was offered a choice of jobs: a resident engineer at a facility in Natchez, Mississippi, a project engineer in Baton Rouge, or a return to the civil design group in Baton Rouge.

As I was weighing the options, it became apparent that a return to the design group was the best choice, and I was transferred back to Baton Rouge in October of 1994. Since my family was in the middle of a school year, it was decided that the move would be made the next summer. This allowed me time to search for a house for the family before the move and still begin my new assignment on short notice. I located a small apartment for rent and began to work once again as a design engineer.

In my new assignment, I reported to another civil engineer whom I had known for almost twenty years. He had also spent time in design as well as construction, and it was an easy transfer. I had been away from design for over five years, and a lot of the procedures and work practices had changed. Most of the design codes had been revised or updated, and I had to become familiar with them again. A lot of the personnel that were there when I left were still there, so I felt comfortable in my surroundings. Some of the people I had worked with were now a part of the other company as a result of the split, and our work was now more like coordinating with a sister company. I also had to become familiar with the changes and additions that had been made at the various plant sites other than the site where I was located in South Carolina. I was assigned as a lead on a new project and began to do the development, preliminary engineering, and field trips required for a complete design.

Some of the projects were located in the plant site in South Carolina, and I was very familiar with the plant personnel and was able to work very closely with them to develop the projects. One project in particular that was assigned to me was a series of environmental cleanout projects mandated by the state EPA. The original responsible engineer for the projects had retired after more than forty years of service. He was assigned to the plant for a six-month duration that lasted for forty years and retired from the site. While I was there during my construction coordination, I was occasionally assigned other projects within the plant. I was registered as a professional engineer in the state of South Carolina and was asked to review and stamp some of the proposed state-EPA-mandated projects upon the retirement of the original responsible engineer. I was then accepted as the new responsible engineer and asked to do additional projects as the need arose even after I had left the plant site.

The environmental projects consisted of collecting all process and stormwater drainage within designated areas and transferring it to a new clarifier and subsequent treatment facilities. Over time, the existing paving had developed cracks in the concrete slabs and joints, and some areas did not have adequate coverage for potential spills and leaks. Most of the collection trenches and sumps were not double-contained, as now required to meet environmental standards, and had to be reworked. These projects, although a high priority, did not make any money for the company but did allow the company to keep operating.

ENGINEERING

On a regular basis, every few weeks, I would fly from Baton Rouge to Columbia, South Carolina, and rent a car to drive to the plant site. My first task was to survey the affected plant areas to determine what was existing and the extent of the required additions or modifications. The engineering department had recently purchased a laser level and a direct reading rod for use in the field with limited personnel resources. Rather than pay for an additional person to fly to the site or to pay for additional labor at the site, it was decided that I could do the area surveys myself.

Existing drawings of the various areas were obtained, and local bench marks were used for the elevation survey. If a known elevation is used as a starting point and a direct reading rod is set on that location, a properly setup laser level will give the elevation of the point the rod is resting on directly from the rod. If the rod is moved to another location and the level of the laser beam hits the position of a moveable target on the direct reading rod, the elevation shown on the rod at that point is the elevation of the new location. By using this method, the laser level is setup and the moveable target is adjusted to the level of the laser beam and the elevation is directly read from the rod. This enables a single operator to set up and determine the elevation of any point desired relative to the known point chosen. If the laser level is moved, a new target location must be determined by moving and locking down the known elevation of the existing point on the rod to match the new. In order to determine the plan dimensions of the area, a simple tape measure can be used and dimensions pulled from the known pipe supports, steel columns, curbed and diked areas, etc. All this information was marked on a copy of the existing drawing, which could be turned into a CAD drawing back at the office complete with elevations and dimensions.

One of the problems that had to be solved was how to determine the elevation of the bottom of a catch basin or invert that was not accessible with the direct reading rod. I visited my friends working for the construction contractor and asked if they could make me a tool from a scrap piece of reinforcing bar. I requested a #3 rebar about five feet long with a ninety-degree bend about six-inches long in one end and the other end sharpened to a point. I also requested that bright-colored tape be placed around the bar at one-foot intervals from the sharp end. The small rebar would fit into any grating or slotted cast-iron cover, and I could use the sharp end to punch through any heavy settlement at the bottom of the catch basin or drain pipe.

By using a tape measure, I could determine the distance to the next foot mark and, therefore, know what the depth was below the top of the grating. By subtracting this distance from the elevation of the top of the grating, I could determine the elevation of the bottom. This could be used every time I visited the plant, but I could not carry it on the plane or check it into my luggage. If I left it in plain view at the site, someone would come along and throw it away as trash. I solved that problem by hiding it. There was a mechanical closet on the second floor of the office and engineering building that had sufficient room behind a large section of ductwork, and the door was never locked. I simply stored my marking tool there each time I left the plant and retrieved it every time I returned, and it was always there.

The solution for the containment areas was fairly simple: replace the damaged or broken concrete with new concrete with proper joint control and water stops for the type chemicals present. If the concrete needed a coating to withstand the chemicals present, it was added to the design. In order to use the existing catch basins and sumps as double containment, each sump was lined with a fabricated stainless-steel liner, and the space between the liner and the existing sump was grouted solid. In order to transport the liquid waste and runoff, it was decided to install steam eductors in the sumps and transfer the liquid to a collector pipe in the pipe rack and send it to a primary clarifier for treatment.

In order to install a piping collector of sufficient size to handle all the local sumps, the pipe racks and supports had to be upgraded. The manifold started out around six inches in diameter but was at least sixteen or eighteen inches in diameter before it reached the clarifier. The initial pipe was supported from T-supports, and they required modifications in order to carry the new loads. Before the

ENGINEERING

manifold emptied into the clarifier, the pipe rack had to have a new level to support it.

This required a lot of field sketches, hand calculations, and computer software calculations to properly size and brace the existing pipe supports for the new loads. Some of the existing catch basins that were out of the affected area were maintained as a part of the stormwater drainage system, but all other catch basins that were not lined for the new sumps were filled in and covered over. It was deemed too costly and time-consuming to try and rehabilitate the existing drainage systems as a double-contained system in accordance with current EPA rules and regulations.

A new clarifier was constructed to handle the updated process sewer flow. In order to construct the new clarifier, sheet pile walls were installed and well points were used to lower the high water table in the mostly sandy soil at the site. The entire project was broken down into several smaller projects and extended over several years. Each individual area had its own problems and challenges that required unique solutions. At the beginning of the projects, most of the plant supervision, especially the operating management was not happy with the project even though they knew it had to be done. They saw it as an interference to their daily operations.

Once the projects were completed, they were happy that their areas had been cleaned up and were much easier to maintain and operate. Looking back, I can say that this was one of the better projects I have worked. It was not because of good weather conditions or comfort because it was sometimes hot, cold, rainy, etc. and I had to wear a hard hat, safety glasses, steel-toed shoes, long pants and shirts, and haul around a lot of survey equipment such as a tripod with laser level, a direct reading rod with target, measuring tapes, prints of drawings, etc. I think the real reason I enjoyed it was because no one was constantly telling me I had to do something different or how to do my job. I understood the requirements, and I simply followed through with the work.

At the end there was a great sense of accomplishment and pride in solving a problem. It also helped when several people told me it looked great or they were very happy with it. There are so many things today that we are required to do but have no sense of pride or accomplishment. Anytime we can complete a job and be told we did a good job, it is worth a lot.

There were also a lot of additions and revisions to the plant sites I had previously designed or constructed. I was greatly aided by having the original design calculations and drawings for the initial plant design available. We still needed to obtain copies of the current plant drawings to account for any revisions made by the plant forces, but the main structures and foundations were generally intact. We used the drawings as a basis to field check any additions and the original calculations, especially the computer calculations, as a basis for modifications for the new additions or revisions. We maintained files in the design office for all the plant sites within the company. This included geotechnical files, stick files for issued drawings, calculations (both hand and computer), specifications, and standard drawings.

By using and building on the existing documents, we were able to respond quickly and provide safe and economical designs for the new plants or plant expansions. Whenever drawings were produced for a plant site, the original documents were returned to the plant for maintaining, and we kept a copy in the design office. If the drawings needed to be revised, a copy was made and left at the plant site while the original was being revised. Since we were now placing all our drawings in CADD format, the original paper or mylar drawings were replaced by electronic copies, but the procedure was the same. Occasionally, a third party would provide a design for the plant for a turnkey project such as a packaging system or an effluent treatment system. We had to consider these designs for space availability or tie-ins for future work.

One of the projects I was assigned involved the replacement of a process tower with a larger diameter tower of advanced materials. A lot of prior planning was required because the plant had to remain in operation far as long as possible before the replacement tower was installed. The existing tower was surrounded by multiple platform levels in an open-air structure. Much of the construction of the foundation modifications and platforms surrounding the tower were completed prior to shutdown of the operation. The tower was supported by a large-spread foundation at grade, and the increased size of the new tower required a larger foundation. The soil was excavated around the existing foundation, steel-reinforcing dowels were drilled and epoxied to the existing foundation, and the bearing area of the foundation was increased to provide adequate support for the new tower and anchor bolts.

Platform modifications were made to accommodate the new tower size. Some steel beams and supports were designed and fabricated but had to wait until after the new tower was set before they could be installed. The new tower was larger in diameter and also had an increased height. This required additional platform levels to service the tower that could not be installed until the tower had been set on the foundation. In order to limit the time required to complete the project, the top two levels were assembled across the street from the site as a modular unit. This unit was to be installed as a separate lift after the tower was set and included all the beams, columns, grating, handrails, etc. for a complete package.

Since this tower replacement was a key element of the overall site plant operation, the timing and scheduling of the work was critical. A team of design engineers were assigned to the site for the duration of the replacement in order to handle and correct any unexpected problems affecting the replacement. I was selected to represent the civil design group for the project. I had worked on the design and was very familiar with what was required and how it had to be accomplished. There were engineering team members there representing project, electrical, instrumentation, mechanical, and civil. I had previously been exposed to modular construction and heavy lifts, but this was on a larger scale than I had known.

The tower was ten feet in diameter and about fifty feet high, and the modular platform was approximately twenty feet by twenty feet and approximately twenty feet high. The crane was set up in a plant road using a main crane and a tailing crane to pick up the tower from a dedicated flatbed trailer transport. The modular platform was lifted from its assembled site along the side of the road using the same main crane for picking and setting the load. Both lifts were made over a process building approximately seventy-five-feet high and set on the opposite side of the building out of sight of the crane operator. Both lifts were made using radio communication between the crane operator and the spotter in order to set the tower and the platform in its correct location. The entire replacement project went smoothly with only a few problems to correct and adjust during the operation. The project was completed, and the plant operation was restarted successfully within the scheduled time allotted.

While in the Baton Rouge design office, I continued to work on projects to be constructed in Louisiana, Texas, South Carolina, Illinois, Arkansas, Kentucky, and Georgia. This included site work, foundations, steel structures, warehouse and utility structures, control rooms, motor control centers, and office space. My time in construction aided me greatly in understanding how the projects would be constructed and how I could make the designs efficient and feasible to build. Each progressive project was built upon the lessons learned from the previous project, and I began to develop preferred methods of design and construction to produce the best safe design possible within the budget and time constraints. By having a basis for these types of design projects, I could concentrate on solving the problems that were different and use the tried-and-true methods for the work that was similar. My work was evolving into more planning, scheduling, and estimating projects and not just the design functions of the job.

A major planned expansion of the company's facilities in southern Arkansas included a new plant to be built in a rural area to take advantage of the nearby facilities and resources. It was to be a similar plant to the facility currently in operation but separated sufficiently such that it would require separate support facilities such as utility, administrative, power, and treatment functions. Our department had extensive knowledge of the facility requirements, and it was a perfect fit for us to design. The foundations were different because of the site location, but everything else was similar.

The layout of the plant was different because of the location. Since it was in a rural location, we had to drill for a water source, and since there was no railway near the new plant, all deliveries and shipments had to be by motor freight, tanker trucks, or pipeline. The geotechnical report indicated a soil subject to degradation due to excessive moisture, and this would prove to be a problem during construction. We had to plan on stabilizing the soil with lime in order to construct foundations at the site. During the design of this plant, I had no idea that it would be the last major design I would do as a part of the internal engineering for the company.

The executive management of the company I worked for decided on a different approach to engineering services. Rather than use internal engineering, a decision was made to outsource the majority of engineering services to a third-party company in order to use these services only when needed and reduce overhead when not needed. This was aided by a Fortune 500 company that had been looking for an internal engineering department. A deal was developed that was beneficial to both companies. My old company was able to divest itself of the majority of the engineering personnel while keeping a few key personnel in specific positions to manage and guide an outside contractor as projects were being developed and implemented. The new company took on the majority of the engineering group personnel and also picked up multiple construction contracts at the old company plant sites.

The new company made offers to the engineering personnel who wished to be a part of the new company including raises, good benefits, and offers for a retirement from the old company when the employees stayed with the new company and met certain minimum requirements. It was a generous offer, and I accepted. As a member of the new company, we continued to work on projects in which we were currently engaged for the old company. On the last working day of December 1999, I shut down my computer, turned off the lights, and left my office for home. On the first working day of January 2000, I returned to the same building, the same office, and logged in to the same computer using the same login and password and began working for the new company.

CHAPTER 17

Owner's Engineering to Contract Engineering

Even though we were still working as a contractor for the former company, our workload was low since finishing our last major project, and I offered my services for things other than design. Within two weeks of joining the new company, I was asked to take a temporary assignment as a contract construction coordinator for the plant design just completed in southern Arkansas. This placed me on-site working as a contractor reporting to the company I had just left and supervising construction personnel for the new company I now worked for who had the construction contract for the site.

Since I was the lead civil design engineer for the project, I was very familiar with it. A few weeks after I showed up, the site construction coordinator left the company, and they sent a construction coordinator from their Houston site to follow the job along with support from me. This construction coordinator was one of my previous supervisors, and we worked together well. We worked the job for a two-week span including the middle weekend. Each of us returned to our homes every two weeks for the weekend, and we alternated so that one of us was always on-site every weekend.

ENGINEERING

The initial site work for the plant was slow because of the weather. It was a wet spring, and the soil had to be stabilized in order to install foundations. We also had a heavy snowfall of around twelve inches that shut construction down and delayed us for several days. The weather conditions also made deliveries of equipment more difficult and time-consuming. After the site foundations were installed and the site grading was reestablished for drainage, it had to be torn up again to install electrical power, grounding, underground utilities, and piping. It was not any different than most construction sites, just more difficult during the wet season. We were able to maintain schedule and the construction work was mostly complete when my services were no longer needed and I was sent back to the office in Baton Rouge to the design group.

Our work was also expanding to facilities outside the US in Europe and the Middle East. My former company was in partnership with a company in the country of Jordan to produce bromine-related products derived from the Dead Sea area and contracted with our design group for preparing the initial preliminary design package for the project.

Although this project involved the usual design functions of site preparation, foundations, concrete and steel structures, office buildings, control rooms, motor control centers, warehouses, and other assorted buildings, it was quite different in the design approach of the preliminary package. Since the project was not in the US, there were different controlling regulations and specifications. The steel and concrete to be used in the project was based on a different specification. All sizes and dimensions were in metric units and had to be placed on the drawings and notes using metric notation.

This is becoming more usual for overseas projects now, but it was the first time we had done a major project of this size in a foreign country. Most of the equipment suppliers were from Europe and the nearby Middle East, but some were from various countries outside of the region. We were initially tasked with providing a preliminary design or FEED (front end engineering design) package. This consisted of providing an overall layout and general arrangement of the project site and the specifications for the project. For civil/structural design, this meant we would show the layout of roads, buildings, structures, types of foundations, the types of steel structures and supports, and the specific civil/structural specifications in order to design and construct the project.

A specification was developed for a geotechnical report in order to design the foundations. Most of the material specifications utilized European DIN specifications. Concrete and structural steel were specified in metric units and sizes. Building design used the *National Building Code* for seismic design and wind loads. Suppliers of materials such as grating came from various countries such as Italy, Germany, Great Britain, and Turkey. Most of the structural steel was based on Great Britain standards and specifications. Many of the specifications and standards that were common for projects within the US had to be developed from other areas. If there was no code or specification available for the project, US standards and specifications were used.

The project was similar to previous projects built in the US, so there was a lot of familiarity with the process, the requirements for equipment support, and the overall plant layout. Our previous design experience was typically for one process plant at a time, and this project involved combining three separate plants into one large facility that would use the same basic raw materials and produce three distinct products at the site. This allowed us to provide support facilities such as maintenance, administrative functions, and utilities with a more economical installation.

There were other unique design requirements for this project. The site was in an earthquake zone that required additional design checks. The site was below sea level and generally very dry, but on occasions, there would be flash floods that would flow through the lower portions of the site without any prior indication of rain. The site was known to have flash floods on days that were completely sunny with no clouds in the sky. The first indication would be the noise of the water rushing onto the site from the mountains to the east. Because of this uncertainty the foundations had to be designed to prevent washout from the flash floods. The soil was also mildly corrosive, and this had to be considered in the design of foundations.

ENGINEERING

Once the various differences were taken into consideration, the project moved forward like any other project.

In the development of the FEED package, there was no detailed design, only preliminary design that was needed to develop the foundation types, size, and structural layout and to develop the overall arrangement. Drawings were produced without specific details or sizes. Approved standards and specifications were provided to develop the detailed design. Each design discipline developed their own standards, specifications, layouts, etc. to be included in the FEED package. The FEED package was issued to the client, and our portion of the project was completed. The client then issued the FEED package to various design firms to bid on the detailed design portion of the project.

Even though the client was a partnership between our previous company and a company in Jordan, they acted as a separate entity for the project. We were not considered for the overall design work because it was thought that there was a conflict of interest because we had developed the FEED package. The detailed design work was issued to a design firm headquartered in Canada. The successful bidder turned around and hired my new company to do the detailed design of the "inside battery limits" (the main core of the project) because of our knowledge of the job. This was a way of getting around the conflict of interest by placing a third-party company in the flow of the work and still using the experience of the group that developed the design package for a maximum benefit.

Once we were given the release to start, we began the process of producing a complete detailed design package that could be issued for construction of the plant site. Even though the foundations would be constructed first, the design of the structures and equipment supports had to be completed to the extent that affected the foundations. Because of our prior knowledge of the plant facilities and requirements, we were able to produce a workable model for each of the separate units, and we began the analysis of the structures. We were held back, as in most jobs, waiting on equipment information for weights and support anchor locations and sizes. However, we were able to predict within a close degree of accuracy. This allowed input of the layouts and support structures into the structural models for analysis. The structural steel members were obtained from steel mills and warehouses that were in metric units and sizes, and these were incorporated into the design models as well as the European specifications.

The one big difference was the plant was located in a seismic zone, and the design had to include checks on the seismic loads in the design of the structures and foundations. The seismic loads were determined from the *Uniform Building Code,* which would later on be incorporated into the *International Building Code* in use today. Extra precautions were taken for design of bracing and connections to satisfy the seismic requirements of the code. The structures were analyzed for the expected modes of failure, and the members and connections were designed to resist those modes of failure with all primary and combined loads anticipated. Once the initial structures and supports were analyzed and properly sized, the anticipated loads on the foundations could be determined. It was several years after our design of the structures was completed, the plant was built. Operations had begun when the area experienced a significant seismic event. There were only minor items that required replacing due to the earthquake. All the major structural steel supports and foundations functioned as designed with no damages.

The natural soil in the area contained a high percentage of sand and granular material and was generally a good material for foundations. Since there was a high probability of flash floods in the area, the foundations had to be supported at a sufficient depth to prevent washout in the event of a flood. Most of the foundations were supported on spread footings, combined footings, or grade beams. The overall site consisted of three major production plants, maintenance support structures, warehouses, pipe supports, load spots, and other assorted structures and supports. The largest plant structure was five levels high including the roof with open sides. Our work within the battery limits consisted of the major plant structures and pipe racks. The design was an interdiscipline effort involving my company's process, piping, mechanical, civil, electrical, and instrumentation groups working together to completely design the inside battery limits facilities including the specifications and drawings that were turned over to the Canadian company responsible for the entire design project.

After the Jordan project was complete, we began working on various sites in the US and abroad. My new company had many projects around the world, and most of our activities were located in the US. We concentrated on petro-chemical plants, gasoline and oil refineries, manufacturing facilities, electrical production facilities (including natural gas, coal, and hydroelectric), nuclear support projects, modular construction, and support of various plant expansions and upgrades.

There was a marked difference in the way we completed our work with the new company. In the previous company, we worked internally to help develop the projects and were a part of the team. We were able to help direct the design effort by suggestions that would produce an economical project that was safe, efficient, and timely. In the new company, we continued to offer suggestions and aid but were at an arm's length to be able to direct the project.

As a contractor, we were hired by the client to provide designs based on their requirements and specifications. Sometimes the client would not have a detailed plan and would rely on us to provide these guidelines. For repeat clients, a good relationship could be established, and our suggestions and their requirements could easily meet for a successful effort. For those clients who did not know what they wanted but presented an offer to bid, we would bid and, if successful, would provide the deliverables as requested. If issues arose during the design and construction phases of the project or if additional work was desired, a change order would be presented to the client requesting additional budget or time to complete. If the client accepted the change order, then the budget and time were adjusted to reflect the change.

Almost all change orders were for increases, but occasionally, a change order would be requested that would increase the time but reduce the budget due to equipment or material deliveries. More likely a change order that shortens the schedule and increases the budget would occur if it saves the client money that is not represented in the awarded work. A good example of this is a request to work overtime to complete a portion of the project ahead of schedule. As a contractor we had to keep a close watch on our budget and schedule because we were limited to time and money based on our contract with the client.

In my previous company, we did almost all the work and only hired an outside consultant or contractor if the work was of a specialty nature or if there was too much work for the internal crew. In the new company, we performed all the work we received or were awarded and added additional personnel if required. Most of the issues dealing with clients were handled by the project team, and once a project was underway, my main concern was to stay within the scope, budget, and schedule and perform the design functions per the specifications, regulations, and project requirements. In both companies, regular project meetings were held with representatives of all parties involved in the project to discuss current progress, planned work for the next several weeks, requests for information with an assigned person, and a date to provide. I was able to work on a larger variety of projects with the new company because of our client diversity, and it helped to broaden my skills and experience.

Some of the major projects included a greenfield power plant built in a rice field, a cogeneration facility installed in an existing structure at a major university in Louisiana, multiple refinery expansions (including a major fire-damaged refinery repair in Texas), methane recovery projects in a landfill in New York, a railcar tank manufacturing facility in Louisiana, completion of a major power plant in Illinois after the original contractor declared bankruptcy, a modular chemical plant constructed in California, and multiple chemical plant expansions in Illinois, Louisiana, and Texas.

One of the more interesting jobs involved work that was truly out of this world. We were contacted by a Swedish firm specializing in a revolutionary type of welding known as pressure welding to provide foundations and structures for a project to support the welding process. The Swedish firm had a contract with a US subcontractor providing fuel tanks for spaceships being sent into space for support of the space station. The structure was to be constructed inside of an enclosed building and would support the sections of the tanks as they were being welded together to form the external fuel tanks for the spaceships. The construction of the tanks was completely enclosed, so there were no wind loads on the structure, only dead and live loads to consider.

This building had been used for previous structures and had many existing foundations, pedestals, and minor structures that had to be removed. The foundations had to be placed around the existing foundations and presented a major challenge. Piles were used to support the loads with massive concrete pile caps to carry the loads from the new structures to the piles. In some cases, the existing piles were used to support the new loads depending on the location. Heads and cylindrical sections of the tanks, approximately twenty feet in diameter, were fabricated at another location and brought to the final assembly structure to be attached starting with the top head, adding subsequent cylindrical sections, and then a bottom head to form the final tanks, approximately eighty feet in length.

As the tank sections were welded together, they were lifted in the structure and a new section was added from the bottom. The structure was open at the bottom along one side for new sections to be added and could be opened up along one side to remove the completed tank. The structure supported the welding mechanism that traveled around the entire periphery of the tank. There were numerous access platforms to support personnel and equipment that had to be attached to the structure. The hardest part was the requirement to limit any deflections to a few thousands of an inch once the welding process had begun. In order to support the structure, we used a set of three high-strength tubular sections connected by braces to form a latticework for each main column section. We started out using the largest commonly available tube section—sixteen inches by sixteen inches with a thickness of one-half inch—and never considered anything less because of the deflection requirements. The columns were supported by massive baseplates and elevation correcting grout for adjustment and support. Almost all the critical connections were welded except the locations where removal was required from time to time. The access platforms, stairs, ladder, etc. were all constructed with bolted connections.

GERALD W. MAYES, PE, RETIRED

I started my career with this company as a senior design specialist working as a lead engineer on most of the projects and became the civil design chief engineer for the design office after a couple of years when my supervisor (civil design chief engineer) left the company to pursue other options. While at this company, my title changed several times and eventually became an engineering manager 4. As I assumed new responsibility, I delegated more responsibility to others in the design group and monitored and reviewed their progress. I became more active in developing scopes, estimates, and schedules for the work. I prepared manpower forecasts and determined if we could handle the proposed work or if we needed to add contract personnel to the group to handle the workload. I was often asked to provide estimates of what my group could do and how long would it take for projects that had not been developed but were still in the very speculative stage.

I learned to estimate manpower and schedules based on past work and performance. This was sort of like looking into the crystal ball for answers, and when the crystal ball was cloudy, I had to pull an answer out of "you know what." My superiors and coworkers did not punish me if I was wrong, but they learned to depend on my experienced guesses since I was pretty close most of the time and better than they were at coming up with an answer. I also learned to rely on my coworkers to provide me information from their area of expertise such as size or weight of equipment, size and height of pipe racks, size of motor control centers, amount of instrument tray in the project, etc. We relied on each other to help develop our own area of manpower and scheduling, and it was great teamwork in pulling it all together.

One of the things we began to do was a process of scheduling projects. It would begin with the key people in each design, management, or support function being notified of the upcoming scheduling meeting in order to prepare for their portion of the work in advance. The meeting would be scheduled for a large conference room with ample wall area to display blank sheets on the walls. Sometimes the major schedule activities would be written on the sheets prior to the meeting. As each area of the schedule was discussed, a sticky note would be added to the sheets to show a specific task. This note would be for a particular discipline or support function and would indicate the task, expected duration, prerequisites for the task to be started, or a time into the task when input would be needed. By placing the sticky notes on the blank sheets, all disciplines could see the interaction between groups and the critical path could be determined.

The critical path was simply the flow of work and duration required by each discipline with interaction between the groups to accomplish the overall work in the minimum amount of time. Any slip of duration of activities or receiving of required information or resources would cause an overall slip in the project completion. Therefore, this path of workflow was critical to the completion of the work to meet the scheduled completion of the project. Once the critical path was determined, any activities that were required to complete the critical tasks that were less than the time shown on the schedule were considered to have float or extra time to complete. This also meant that tasks that had float could be accomplished in a longer time span without affecting the critical path and could utilize less manpower to complete.

Likewise, any task on the critical path could be shortened by increasing the manpower allocated or by utilizing overtime to reduce the total time required. This could result in another path becoming critical or by shortening the overall schedule. The use of the blank sheets with moveable tasks and durations allowed the team to discuss and formulate the best possible schedule for the project. After the schedule was approved by those present in the meeting, a picture was taken of the sheets on the wall, and the project scheduler would then transfer the information to a formal schedule to be reviewed again for agreement before it was published as the official project schedule.

As my knowledge and experience grew, I was able to participate in company-wide endeavors that involved travel and cooperation with my peers in other offices and parts of the company. I was involved in team efforts to develop practical methods and procedures to perform the design functions in a more efficient manner. I was also able to gain knowledge from others that aided me in my planning for future projects. I spent a great amount of time developing standards and guidelines for the design office to be used in future projects.

One of my assignments was on a peer-review team for a project to be built in Saudi Arabia. Our design office in Great Britain was tasked with the design of a large chemical facility, and a team of knowledgeable engineers were assembled to review the progress and adherence to codes, guidelines, and standards. We were flown to Great Britain, and for five straight days, we left the hotel and walked to the office (about ten blocks away) and reviewed drawings, guidelines, specifications, standards, schedule, and conformance with project requirements. We then walked back to the hotel after a full day. Lunch was provided at the office, so there was no reason to leave during the day. This was in December, so we arrived in the dark and left in the dark. As soon as we finished the last day, we headed for the airport to return home. This was a chance for me to see how other design teams handled large projects and was very enlightening.

CHAPTER 18

Additional Experience as a Critical Lift Specialist

One day a request came through wanting to know if anyone had experience with critical lifts. During my time in construction, I had been involved with critical lifts for various structures and equipment, and I indicated an interest in developing more knowledge on the subject. The need arose because the company critical lift specialist was retiring and his understudy had not yet acquired his professional engineering license. I had a regular job in the design office, and this assignment was to be temporary and on an as-needed basis.

I was sent to Denver, Colorado, to meet the retiring specialist and become familiar with the company practice. Per the company policy, all lifts over a certain weight and any lifts to be made over personnel in buildings or work areas was considered a critical lift. The company policy provided a procedure for developing a lift plan, who was to provide input and review, and what degree of safety was to be followed in the equipment, tools, and activities related to critical lifts. Normal construction activities involving lifts that did not rise to the level of a critical lift were covered under construction guidelines.

If the proposed lift was considered a critical lift, the initial lift plan was started by the person in charge of the lift, usually someone that handled lifts for the construction contractor. The company procedure for critical lifts was followed, and the plan was submitted to the company critical lift specialist for review. The plan would consist of a sketch showing the location of the equipment or material to be lifted before and after the lift and the position of the lifting crane. Any obstacles would be shown in the sketch both horizontally and vertically. The position of the crane would indicate the area of mats and ground-bearing capacity beneath the crane outriggers or support track. If there was to be movement of the lifting cranes, the movement would have to be described or shown in the sketches including the swing or reach of the lifting crane to be used. The data for the lifting crane would need to be specified so that proposed lift could be verified as being within the safe range of the capacity of the crane.

All lifting equipment utilized below the lifting hook of the crane had to be specified for capacity, and the required safety factor had to equal or exceed the requirement for the particular type of lifting equipment. If a piece of lifting equipment was to be designed specifically for the lift, such as a spreader bar, it had to have been designed and stamped by a civil/structural engineer and documentation had to be provided as a part of the lift package. I had designed and provided calculations and verification of lifting components previously, but now I was being asked to review and approve a lifting plan provided by someone else. All lifting plans were to be submitted at least one week in advance for approval, and on major lifts, additional time would be required.

Some lifting plans were simple and could be reviewed and approved in a matter of hours while others required several weeks to complete. All equipment used had to be inspected and deemed worthy before it could be used. This varied from a visual inspection of slings and clevises to required certification of lifting cranes that required detailed annual inspections and visual inspections before each use. Cranes are listed with maximum capacities for the shortest boom and reach configurations and will differ when lifting over the front of the crane or when lifting over the side of the crane. As the

ENGINEERING

crane boom is extended and the reach is increased, the safe lifting capacity of the crane is reduced. It is the job of the person initiating the lift plan and the person reviewing and approving the lift plan to ensure that the reductions due to the lifting arrangement are considered and the lifting plan proposed is within the safe limits of the crane and equipment used.

I served as a lift specialist for the company for a period of a couple of years and saw a wide variety of lifting plans. Shortly after I became the backup for reviewing lift plans, I was asked to visit a site in Victoria, Texas, to review a lift plan. The person that covered the most of the regular lifting plans for the company had to be out the week the major lift was planned, and I was asked to provide a review and approval of the plan.

The lift consisted of a large vertical vessel that had to be lifted vertically and then moved horizontally before setting in place on a concrete foundation with anchor bolts aligned with the baseplate flange of the equipment. The total weight to be lifted was between 550 and 600 tons and was definitely considered a critical lift. In order to lift the equipment, a large crawler crane with added weights for balancing the load was used. To help lift the vessel from its horizontal position on the large trailer that delivered the vessel to the vertical position, a secondary tailing crane was used to initially lift the vessel from the trailer and assist in putting the vessel in the upright position. To transport the crane, the boom, additional weights, and the tailing crane to the job site, a total of ten eighteen-wheeler transport trucks were used for the individual parts required for the completed crane.

To assemble the crane, an entire block of a plant road was barricaded off, and then the boom was assembled and the cables run and attached to the crane. I had worked a full week with the other lift specialist in reviewing and approving of the lift plan, and after he left, an additional week was required to complete the plan and provide final approval. The crane had been inspected and met the requirements for a safe lift in the proposed configuration presented in the critical lift plan. The tailing crane was satisfactory for a safe lift. All slings, spreader bars, and lifting equipment were inspected and had documentation indicating a safety factor in excess of that required for a safe lift.

The actual lift was made on Saturday around noon of the second week. All lifting procedures were accomplished per the lift plan, and the vessel was set on the foundation and secured in place with the anchor bolts. A slight wind developed during the lift, but it was not enough to halt the setting of the vessel. My job was finished, and I returned home to Baton Rouge.

Over the next couple of years, I worked on various projects containing critical lifts in Kentucky, Texas, and Louisiana. Lifts were made for surge protection gates to be installed near New Orleans following Hurricane Katrina, equipment installations in oil refineries in Kentucky, equipment installations in petro-chemical plants in Texas and Louisiana, and for various fabrication facilities located in South Louisiana supporting the petro-chemical industry and the nuclear industry.

Most of the lifts at the fabrication facilities were moving equipment or structures from the fabrication site to truck transport or to barges for transportation to the plant site. Most of the lifts at the final site were moving equipment from either train cars or transport trucks to temporary storage locations or to the final position supported by steel structures or concrete foundations. In almost all the lifts there was a component of load-bearing capacity of the soil or paving at the location of the crane setup that needed to be addressed.

This is especially appropriate for a civil engineer since they are knowledgeable about soil capacities. Some people functioning as a lift specialist have a background in mechanical engineering or field experience with the mechanical functions of the lift but are not experienced with the soil mechanics functions. There were cases that involved transporting heavy loads by crane or truck across underground conduits and culverts that had to be analyzed to determine if they had adequate capacity. In some cases, a number of structural members had to be temporarily removed to set the equipment in place and replaced after the equipment was installed. In other cases, a detailed analysis of the supporting structure was required before the lift could be completed.

ENGINEERING

In each case, the lifting crane had to be evaluated for the loads imposed in the various lifting configurations and the appropriate safety factor had to be verified. In the instances in which the person initiating the lift plan did not adequately address all the requirements of the lifting plan, the lifting specialist would follow through with the appropriate information or documentation. Most of the requirements were satisfied by obtaining the crane data, the lifting components documentation, or by providing a detailed sketch of the lifting plan and sequence of lifting to complete the lifting plan requirements.

One of the things I learned while working as a critical lift specialist was the cause of most lifting failures and disasters. The large majority of failures were directly attributed to operator or human error. Most companies have elaborate lift policies and guidelines when handling critical lifts. All the components of the lift including lifting lugs, shackles, slings, spreader bars, and cranes have to be either certified, tested, or designed by a competent engineer who is responsible for making sure the components meet or exceed the safety factors required by the various lifting codes and guidelines. Usually, there is a written description of the lifting procedure that is reviewed and approved prior to the lift.

Field conditions must be considered such as overhead or nearby obstructions, the planned route of the lift from delivery to the site, off-loading, lifting of the load using the main lifting crane and the tailing crane if required, horizontal movement or travel of the load, and setting in place of the load on the foundation or support. The lifting and tailing cranes must be inspected for damage and serviceability prior to use. The lifting hook and cables, which might be transported separately and installed on-site, must be inspected for damage during the installation of the crane components prior to the lift. The area beneath the lifting crane must be assessed to determine if it will support the designed lifting pressures from the crane while making the lift.

If the area beneath the crane is not adequate to support the loads, it must be reinforced with structural backfill or timber mats may be used to spread the load. In some cases, the ground beneath the route of the transported load may need to be investigated and reinforced. Weather conditions may need to be considered such as high winds that could make the lift unsafe. All these things must be investigated and reviewed before a critical lift can proceed.

Over the years, I have taken an interest in all the reported lift failures, and most all of them can be traced back to someone not following the accepted guidelines and written procedures for a safe lift. Of course, there are cases of equipment failure that could not have been detected by regular inspections, but these are few and far between.

After a couple of years, the lift specialist position was filled by a rigging engineer who had many years of experience with lifting plans, and I phased out of that type of work. However, the individual taking my place was not a civil engineer and needed help from time to time to review and check soil-bearing capacity issues. I continued to support the critical lift specialist group as needed for several years. I attended additional training classes on rigging and critical lift to remain current.

Anytime a critical lift was a part of the design function, I was able to lend my expertise to the group. This involved the location and staging of lifting equipment and transports as well as lifting lug analysis and design. It also proved beneficial whenever clients needed equipment removed and replaced in structures that were not originally designed to be removed. Many times, temporary supports and braces were required in order to safely remove and replace equipment in structures that were in service and providing supports to operating equipment. We tried to provide designs that were capable of handling changes in the future with as little as possible interruption to the operation of the plants.

CHAPTER 19

Consulting for the Construction Group

Our company had a very active construction group that was involved in the projects we designed as well as other projects all over the country. From time to time, we were requested to provide reviews and evaluations of construction projects to support the construction effort. I was often called to support the construction effort because of my past experience and knowledge. I had spent over five years for my previous company in construction coordination and had also been contracted out as a construction coordinator for about three months when I first arrived at my present company. I had been involved with explosion and fire-damaged rebuild projects on multiple occasions with my previous and current companies. I had experience with critical lifts, structural inspections, site evaluations, and providing solutions to various field problems.

On multiple occasions I was asked to provide knowledge-based expertise for site selection of new plants as well as undeveloped areas of existing plant sites. If time allowed, there was usually a soil investigation conducted of the area to determine the existing soil conditions, the feasibility of using the existing soil, modifications of the existing soil, requirements for bringing in new or additional soil, and the type of foundations required to support the anticipated loads. This could be a significant cost and delay to any project and needed to be evaluated early in the process. Even if the physical plant was identical to a previous plant, it could be significantly affected by the foundation requirements.

If the proposed site was to be located in a remote or distant location from other similar plant sites, several other items had to be considered by the selection team. What is the availability of rail and roadway access to the site? How are raw products received at the site, and how are finished products transported from the site (roadway, rail, container ship or barge, pipeline, etc.)? What is the availability of manpower to operate and maintain the site? Is there sufficient electrical power capacity in the area? Are there sufficient utilities in the area (water supply, wastewater treatment, and nearby pipelines)?

If the site had been selected, our group was often asked to do a design for a preliminary estimate to include foundations, structures, diked areas, paved areas, and various support buildings such as electrical substations, control rooms, office buildings, maintenance buildings, warehouse facilities, and stores and receiving buildings. A knowledge of the building use was very helpful to determine the size required. Were the control room and office buildings to be used for continuous operations or just during a single day shift? Were the stores and receiving buildings expected to handle a large size and volume of equipment and parts or a limited amount? Will the plant require a large amount of continuous maintenance and multiple operators or can it be operated with a small crew utilizing automated controls with limited maintenance? All these factors were taken into consideration when providing a preliminary design basis for an estimate. As my knowledge and experience increased, I was able to provide better preliminary designs, and this resulted in more accurate estimates.

Another area that I was involved in was the evaluation of existing structures for modification or additions. The decision to rework or modify an existing building or structure often needed to be evaluated during the scope development of the project. This usually involved visiting the site for an inspection, gathering of existing drawings and site data, and returning to the office to evaluate the existing structure or building for the proposed loads. It was important for the scope of the project to determine if the existing structural members and foundations were sufficient to handle the proposed loads or if major modifications or upgrade were required for the project.

This evaluation often determined if the proposed site was adequate or if a new site and location would be needed. This was not an absolute determination but was enough to provide a high probability of the success or failure of the proposed location. Since most loads were approximate, the structures and buildings would need to have a final evaluation using the correct loads to determine if the structure was adequate or if minor modifications were required.

Even though most of the scope was well-defined at the start of the project, there were always some unexpected changes to the project, and I was sent to the field to handle unforeseen problems during the construction phase. The largest area of concern was the underground. Anything you could not see was likely to be lurking beneath the surface and was usually in the location that gave the most problems. One of my favorite items was underground electrical duct bank. It was always shown in a nice neat location on the drawings, but the electrical installation contractor did not always follow the drawings but rather made wide sweeping curves from point A to point B. The location could only be accurately shown if someone did an as-built location drawing (which was almost never done), brought in a third-party contractor to run belowground radar to determine what was underground, or try to install a foundation and discover the duct bank interference.

Either way, the duct bank location had to be avoided and usually meant a redesign of the foundation that interfered with the duct bank would be required. Other assorted interferences included foundations poured too large underground or not in the correct location, underground drain lines that may not have followed the locations shown on drawings, and firewater lines underground. There was also the case of underground interferences that were not placed on location plans or the location plans showing the underground obstacles were not able to be located during the existing drawing research.

One of the large areas for conflict and confusion is the placement of pipe, conduit, cable tray, and instruments in overhead pipe racks. Since this is an area used by multiple disciplines, it is necessary for each discipline to convey their designs and locations to all other disciplines using the pipe racks. On a well-planned and coordinated project, this will be done by each discipline. But occasionally some overlaps will occur, and adjustments must be made. Usually at the start of the project, the existing pipe racks will be surveyed to determine location and use of the rack so that interferences will be kept to a minimum.

However, some projects may be using the same existing rack as a similar project with another design company at the same time, and if proper coordination is not maintained, there will be interferences. Also, small plant projects may install pipe or conduit in the existing racks that interferes with the current project. These interferences can usually be minimized using proper coordination between all groups. There can be interferences between piping, cable tray, and conduit that has been marked for demolition and removed from the drawings but still exist in the field.

On some projects, members of the design team were asked to go to the construction site to assist construction forces during a critical turnaround or construction effort. See chapter 16 for the tower installation located in the Arkansas plant for the previous company. It was thought that if any issues came up during the intense construction effort, the design engineers familiar with the project and the specific design could supply an immediate solution if on-site. This also aided in the accurate flow of information and the understanding of the problem.

In some cases, the problem was solved immediately in the field, and I returned to the office to correct the software analysis and drawings at a later date without delaying the construction effort. If the problem could not be solved immediately, a call to the home office to ask another engineer to rerun the analysis program with the adjusted data tended to speed up the solution. If all else failed, I could return to the office and work on possible solutions to the problem having already observed the problem with a full understanding, which saved time and effort in the final solution.

In addition, problems encountered in the field were much easier to visualize and understand than trying to solve problems from an e-mail or a marked-up drawing. I believe one of the major problems of young engineers and designers is the lack of field experience in being able to understand and visualize the problems encountered in the field when a project is being constructed and when it is to be placed in operation in a safe and efficient manner.

One of the more interesting projects I was involved with started out while I was walking down the hall passing the office of my boss. I was asked if I knew anything about mass concrete. I had knowledge of mass concrete resulting from large mat pours I had designed and was able to observe the installation of them during my years in design and construction. I replied that I had knowledge of the requirements of designing and constructing mass concrete installations, and I was invited into their office. This began my involvement in a project that I helped secure for my company based on my knowledge of mass concrete.

The project was a modified hydroelectric plant located in North Carolina. I am not sure what happened to the previous construction contractor. They either went bankrupt or were not able to satisfy the requirements of the project and the major power client. The existing plant was located adjacent to a large-capacity lake with an earthen levee. Water was control released from the lake and flowed through turbines at the base of the levee, supplying power to generators that produced electricity. Because of updated regulations and concern over possible seismic events in the area, the levee was scheduled to be reinforced with a larger base.

To increase the width of the base and decrease the slope on the downside of the levee, the existing power plant had to be relocated downstream from its present location. This also allowed the power plant to be modernized with increased efficiency and capacity. The design of the power plant and levee reinforcement was already finished and the construction had begun when the client began to seek another construction contractor to complete the work. I just happened to be in the right place at the right time to provide assurance that we were capable of providing support to our construction group for correct installation of massive concrete pours.

The American Concrete Institute (ACI) defines a *mass pour* as "Any volume of structural concrete in which a combination of: dimensions of the member being cast, the boundary conditions, the characteristics of the concrete mixture, and the ambient conditions can lead to undesirable thermal stresses, cracking, deleterious chemical reactions, or reduction in the long-term strength." When large continuous pours of concrete that exceed three to four feet in thickness are planned, they should be further evaluated to determine if they are classified as a mass concrete pour. We were dealing with pours in the range of five- to six-feet thick with a total thickness of around twenty-five feet in some places.

I began my research in all the controlling regulations and codes to make sure we were complying with all requirements. When the cement, large aggregate, small aggregate, water, and any approved additives are combined to constitute concrete, an exothermic chemical reaction takes place that produces heat. In a massive installation, the buildup of heat is detrimental to the proper curing of the concrete. To the casual observer, it would seem that the problem with mass pour of concrete would simply need to be cooled. It is not that simple.

A larger problem is the heat buildup in the center of the concrete combined with the cooling of the concrete near the edges. This will allow the concrete along the outside edges to contract during the cooling process and produce cracks in the concrete, which is very undesirable and can affect the capacity of the pour if the cracks are deep enough. If the cracks extend to the reinforcing, the concrete cover protecting the reinforcing is lost and the reinforcing can be damaged. Also, significant cracks can lead to water intrusion, which will further affect the concrete.

Concrete additives are often added to the mix to help control the rate of curing and, therefore, help to slow the heating center of the pour. The water used in the mix is cooled before adding to

ENGINEERING

reduce the pour temperature. Sometimes pipes are encased in the concrete installation so cooling water can be circulated through the pour during the curing process. In order to slow the cooling of the edges of the pour, insulating blankets are used to hold the heat in the pour. Whatever method is used, it is important to attempt to keep the temperature at the middle of the pour no more than 160 degrees F and no more than 35 degrees F greater than the temperature at the edges. These efforts to minimize the temperature differential may take several days to accomplish.

For our particular project, the pours were installed with PVC pipes running through the middle of the pour that could deliver cooling water on a continuous basis. After the pour was made and the surface hardened, insulating blankets were placed on the edges and the tops of the concrete. The bottoms of the pours were either on natural grade or on a previous pour that provided an insulating effect to the concrete surface. Temperatures were monitored, and when they fell below the accepted levels, the insulating blankets were removed. The flow of cooling water through the concrete was discontinued. The PVC pipes were grouted with high-strength grout to maintain the integrity of the concrete installation.

After the first pour had cured and was capable of accepting a load, the new pour was placed on top of the previous pour, and the installation process was repeated for each layer. Some of the layers were designed to allow water to flow through channels on the way to the turbines, and others were designed to support various pieces of equipment including the generators. Each layer had a different reinforcing layout, and reinforcing bars were extended out of the pours to allow for proper splice lengths to the adjoining sections.

For each section, the PVC pipes had to be installed around the reinforcing and any concrete block outs in the section. Each mass concrete pour went smoothly, and the construction personnel and supervision followed the guidelines and closely monitored each pour. We also found a mass concrete expert in our Canada office who reviewed and concurred with our recommendations for the installation of the concrete.

During this project, I made several trips to the site to review the construction procedure and was asked to provide additional support for other construction issues. To install the concrete supports and equipment foundations along the base of the existing levee, retaining walls had to be installed in order to hold back the existing levee and surrounding soil. I was asked to review and provide guidance with the existing retaining walls installed by the previous contractor and any additional retaining walls required as the installation progressed. I discovered while I was there that no one was keeping the client informed regularly of the progress of the retaining walls and the mass concrete pours and requested changes in the design required because of site conditions. The client was often being informed of what had happened rather than discussing with them what was requested and why it was needed.

While at the site I visited regularly with the client and discussed these matters with them. They were very supportive of what was being done once they knew the reasons and had input into the decisions. During this period, it was noticed that an elevator was shown on the construction drawings with capacities and levels to be serviced but was not fully specified. No one was working on this, and it was becoming a crucial item because of the concrete to be poured around the elevator shaft. I asked permission to research and evaluate an elevator contractor to provide and install at the site. I found a supplier and met with them in their office and explained our requirements. They were able to provide typical installation details and requirements and clearances for the elevator installation.

I carried these back and discussed with the job site construction manager and the client. This vendor was approved and a purchase order was issued. The elevator was installed by the vendor with no problems. Even though I had never specified an elevator before, it seemed the logical thing to do since I needed the size of the elevator shaft, the clearances required, and locations of supports to complete the concrete installation and no one else was available for the task.

During this time of multiple trips to the field, I still had a civil/structural group reporting to me back in Baton Rouge. Work was slow in the office, which was one reason I was able to follow the construction project. There were many things that needed to be done at the site with no one to handle. I began, with permission, to bring work back to the main office. Engineers and designers working for me were able to verify changes needed in the design drawings to fit the site requirements. Material takeoffs were needed, and our group was able to handle with ease.

I became so involved with the construction effort that I was asked if I could transfer to the site as the civil site engineer. I enjoyed the work but respectfully declined because I had people in Baton Rouge that reported to me and I was responsible for them. The construction effort was completed successfully, and I was glad that I was able to support them with my experience and knowledge.

Having worked for a petro-chemical company in my past employment, I was very familiar with fires and explosions that interrupted the various facilities productions. Most people think that chemical plants are very dangerous and would not want to work at one, but the opposite is actually true. Not all but most of the chemical plants I have knowledge of were very safety conscious and good places to work. They all have very strict rules and procedures to ensure the safety of employees and the communities that surround them. Just like the general population, there are accidents and mishaps, but the chemical companies take extreme measures to control and alleviate any instances that arise. If most residences and businesses took the same precautions that the plants did, there would be much less accidents and issues overall. The chemical plants do, however, handle much larger quantities of hazardous materials and need to be aware of potential problems that could affect many others.

During my career, I have been sent to the field in Texas to provide civil/structural design for the rebuild of a chemical plant that had an explosion during a turnaround start-up. I was in the field for two months assisting construction with on-site design. I was in the field as a construction coordinator in South Carolina when an explosion occurred in another plant. It was decided that the rebuild effort would be handled by plan forces and contractors familiar with the site, and I was designated as the person in charge of the construction effort. Contract construction coordinators reporting to me were utilized in the effort as well as continuing the ongoing construction effort at the site.

I brought in a contractor to handle on-site fabrication of steel and piping on a night shift and assigned different duties to each available contractor during the day shift and coordinated all efforts with plant operations and maintenance to complete the effort in the shortest time practical. I set up a small trailer in the middle of the work area with no bathroom or air conditioning. The weather was well over one hundred degrees each day, and I did not want the construction forces and supervision to have a reason the stay in the trailer any longer than necessary for meetings. We held a meeting once a day to check on progress, planned activities, and any problems. Tasks were assigned, and everyone was quickly sent back to work to complete the rebuild.

Because of our experience and knowledge of fires and explosions, we were often hired to provide design or construction assistance to other companies after they experienced production interruptions due to fire or explosion incidences. At my present company, we were requested to support a refinery in West Texas that had experienced a major fire at its facility. A temporary facility near the refinery was secured and was staffed with personnel from our Houston, Texas, office as well as our Baton Rouge office. I was asked to provide a field group of civil/structural engineers and designers to provide design assistance to the reconstruction effort. The personnel worked for me but, for purposes of the field effort, reported to a project manager who was in the field continuously.

I made several trips to the field and provided design sketches, review, and coordination for the duration of the effort. We also were called upon to provide design and construction assistance for a plant explosion at a plant site south of Baton Rouge. Personnel reporting to me were sent to the field to gather information, provide drawings, and any construction assistance that was requested. I was involved in the effort to review, issue drawings, and provide consultation but was not sent to the field because of the other work being handled at the same time in the office.

ENGINEERING

During my career I became a source of knowledge and experience for construction and field problems. This grew out of my time spent in design, on-site start-ups, construction coordination in the field, and troubleshooting problems that arose. Most of the experienced people I dealt with were mostly in one main area, such as design or construction, but I had experience in both areas as well as supervising crews in design and construction. I had been on projects in which I had to solve problems in real time to keep things moving along and was not afraid to take charge and make decisions. This helped me to advise the construction personnel and provide realistic solutions even while in the design office. I was always happy to offer my opinion, and even though my advice was not always able to be followed, it was appreciated as a possible solution and was sought regularly.

CHAPTER 20

—m—

Modular Design Experience

My first experience with modular design came when I was a construction coordinator for a plant site in South Carolina. I was previously familiar with the concept of modular design as related to residential and commercial construction. Sections of wall were prefabricated and installed as a single unit. Ceiling joists, rafters, and bracing were incorporated into prefabricated roof trusses and installed as a single piece to aid in time and quality of the constructed items. Roof trusses could be prefabricated using template and jigs faster and with more quality control than the individual lumber members could be constructed in the field. It saved time for assembly and erection and reduced the cost of the trusses.

Now I was seeing what modular construction could do for the petro-chemical industry. Equipment began to arrive at the site that was prefabricated with as much assembly as practical. I had viewed equipment skids in the design office while checking approval drawings previously, but I now began to see the huge advantage at the construction site for erecting modular designed equipment. Equipment skids arrived with the equipment such as a small drum and a pump that was assembled with connected piping, valves, instrumentation, steam or electrical heat tracing, selected insulation, and support

ENGINEERING

steel and platform walkways (including grating) fully connected and supported.

The equipment skids were often painted and had already been pressure tested and ready for service. The field forces simply had to off-load from the transportation source and install at the designated field location, which was usually a foundation with anchor bolts or a location on an existing structural member that could be either concrete or structural steel. The connections of piping, instrumentation, heat tracing, etc. were made, and final insulation or painting completed the installation. This allowed for a quick installation in the field and the assurance that the items on the skid were accurate and had been tested before leaving the fabrication shop.

Without the use of modular construction, the field forces would have been required to erect a portion of the supporting steel framework and wait for the delivery at the site of the equipment. The equipment would then be installed in the skid, and the piping and instrumentation would have to be added by another craft in the field. Pressure testing of the pipe system would be required before the piping could receive its final paint coat or insulation. The ironworker crew might have to return to add the final platforms, gratings, or other supports to complete the installation. Even in the best of conditions, the total installation might be delayed waiting on some of the items to arrive.

By prefabricating the skid in the shop, all items are received, assembled, and tested before shipping to the field for installation. This allows a shorter period to accomplish the construction effort, which is often a premium, especially if the installation has to occur during a plant shutdown. By using modular construction, all items can be procured, fabricated, installed, and tested, allowing the installation to occur during the short window of opportunity.

The use of modular construction is not only used for equipment skids but for anything that can be preassembled and installed as a complete unit in the field. This has always been used for ease of installation for things such as pipe bridges, pipe bents, precast catch basins, pipe spools with installed valves and fittings, etc. But when items from different crafts can be preassembled for ease of erection, it speeds up the construction effort and provides a high-quality installation.

A few years later I was sent to the field in Arkansas from my design office in Baton Rouge to follow a construction effort as a design backup if needed (see the previous reference in chapter 16). The timing for completing the turnaround was tight, and the existing structure had to be modified in order to accept a larger piece of equipment. As much modification as possible was completed prior to the plant shutdown. Some plant shutdowns only involve a portion of the plant site production, but others will shut down the entire site unless some preshutdown storage of product allows a part of the site to remain in operation. This project involved the entire site and was scheduled to be shut down, modified or repaired, new equipment installed, and the plant brought back into production (hence the term *turnaround*).

In order to complete the new installation while removing and modifying the old structure and equipment, the entire top couple of floors for the new installation were being constructed across the street from the final location. This involved structural steel, equipment, piping, etc. and was all preassembled. The original area was demolished and modified at the same time the new structure and equipment were being assembled. The original vessel was removed and replaced with a larger vessel. Once the new vessel was in place, the entire structure complete with assembled equipment, piping, etc. was lifted as a modular unit, swung into place, and attached to the existing structure, which had been modified to accept the new top section. The lift and installation of the modular section proceeded as planned, and the plant was able to start up within the designated time period. The entire plant was back to full production. I also added to my lift experience that had begun when I was in construction.

I continued to gain experience in modular construction as it added to the efficiency of any construction effort when it was practical. The company I now worked for utilized modular construction whenever possible, so there was a lot of need for modular design. For the design projects we developed, there became a set of rules or guidelines to be followed for transporting modular designs, and these were incorporated into the layouts and arrangements. If we were designing a set of pipe racks, pipe bridges, support platforms, etc., it became important to separate the modular units for maximum efficiency.

For smaller projects or remote locations, we had to consider the route to be taken for the highway transportation to reduce cost and complication of moving along the highways, especially two-lane roads near the plant site. If we were transporting along the interstate system, we had to consider the highway vertical clearance beneath the overpasses. Even on a lowboy, the height of the lowboy bed had to be added to the height of the modular unit being shipped. If the height was more than allowed under some overpasses, then the trucks would have to exit the interstate and go around any low overpasses whether it was a few hundred feet or a few miles. We always tried to avoid this if possible.

If the trucking company placed dunnage or timbers beneath the modular unit, it also added to the height of the load. There are also restrictions on the width of load that is allowed to be moved over the interstate or over local roads. If the load exceeds the width of the trailer, it must be marked as a wide load. It may require an escort to travel with the load to identify it as a potential hazard so other traffic can be aware of the shipment. If the load exceeds a certain width, then the transportation can only be made during daylight hours, and if the distance is great, it may take several days to arrive, which adds to the cost of the shipment.

To be absolutely sure of which requirement has to be met, the route must be determined and the appropriate authorities must be contacted to assure a problem-free passage can be made. Even though there is no absolute guide, in general, the modular load should not be more than eleven feet in height or width including any connection plates, gussets, or extensions in the height or width of load. This also includes equipment nozzles, piping spools, instrumentation, insulation, etc. The secret was to make the most of the allowed dimensions while breaking the modular sections at the appropriate locations that would make the most sense for the installation in the field. This was a defining item when laying out the modules to prevent problems at the end of the design phase when everything was set to the layout first provided.

For instance, if the layout of a platform clears all the above-mentioned parameters, but when during final assembly check, you realize that the stairs or handrail that is a part of the module stick out past the outside structural member by six inches and it prevents you from shipping because of height or width restriction, you would not cut the six inch off and reattach in the field. You have to have a vision all the way through the design that takes into consideration these things and make allowances. When piping gets installed in the module, it often extends beyond the edge of a column or beam in order to have sufficient distance to make a field weld or connection, and these things must be considered.

Because of our group's experience, we began to get requests for design and layout of modular sections for projects that were not being designed by others in our office but by other offices within our company. This provided a steady stream of work and increased our knowledge and proficiency in these types of projects. Our company provided supports and installation for equipment and related piping, electrical, instrumentation, and associated disciplines. These clients included major chemical plants, electrical generation and distribution facilities, manufacturing plants, nuclear facilities, refineries, and other large industrial clients. Because of the size of the projects and the need to complete them on time and within budget, the use of modular design and installation became a necessity. The cost of projects running hundreds of thousands and sometimes millions of dollars per day during critical periods dictated that the work be accomplished with the maximum efficiency in the shortest period of time.

Modular design and construction are best suited for this, especially during peak periods of construction. The fabrication for modules could begin in the shop at the same time as the site work is beginning in the field and can be delivered at the site in a timely manner to be installed by the heavy equipment and cranes at the site. The scheduling of equipment and modules in the field has to be tightly controlled so that there will be available crane access. To reduce the cost of the crane on-site, a carefully planned schedule is needed to utilize the crane effectively. The use of modular design and installation helps to bring this about.

One of the items using modular sections is the many pipe racks and pipe bridges required for most industrial plants. These racks carry the process fluids in pipes, electrical cables in conduit and cable tray, utility piping for the plant steam, and cooling water, flare lines, and sometimes supports maintenance platforms and coolers (fin-fans). The supports are typically placed at a nominal distance from around fifteen feet to twenty-five feet apart depending on the ability of the piping supported to span the distance.

A typical span would be twenty feet and would consist of an H-shaped structure with vertical structural members called columns and multiple horizontal transverse structural members as needed. These H-shaped structures are called bents and are usually connected to each other with structural members called stringers located at the midpoint elevation between the transverse members or beams. A typical spacing of the beams is from three feet to five feet apart with the stringers at the midpoint of the beams located below the top level of the beams and can be as many or as few as needed. For small pipe racks, the bents may be fabricated in the shop and delivered to the site for installation on concrete foundations, followed by installation of the stringers in the field. For larger pipe racks and pipe bridges, a length of the connected bents or the total bridge length (if it can be transported as a unit) can be shipped as a module and connected to the pipe columns or pipe bridge columns in the field. This method was successfully used for a major power plant expansion located in the state of Maryland. I also remember a project that was installed in the Middle East, which I did not design but participated in a peer review of the design, that was installed by a large crane with single bents approximately fifty-feet wide and over sixty-feet high.

We were asked to do a modular design for equipment that would be placed in an existing landfill site in the state of New York. The purpose of the project was to reclaim the methane gas being created in the landfill, clean and purify the gas, and place it into a natural gas pipeline near the site. Since the landfill site was continuing to decompose and compress, pile foundations were required to support the equipment. All the equipment, structures, tanks, etc. were placed on modular skids and installed on the pile foundations for support. This allowed the top level of the soil at the site to consolidate without affecting any of the site equipment and structures. It was a very successful project in design and was proceeding on schedule for installation until the construction was halted due to unforeseen circumstances. The landfill that had been closed was reopened for the use of disposal from the debris from the World Trade Centers in New York that were destroyed during the 9-11-2001 attack. The project was completed at a later date.

Modular design and construction have proved to be very valuable tools for large as well as small industrial or commercial sites. If I really want to be honest, the first time I used modular design and construction was when I was a kid playing with my Tinkertoys (my generation's version of LEGO toys). I realized that when I wanted to make something different, I only had to disconnect the pieces that were necessary to construct a different model. This was my form of using the modular design to construct my finished model. For instance, I would not have disconnected a set of wheels from the axle if I was going to use the wheels again but would start from the assembled axle with wheels attached.

Often in modular design, if there is a problem with the size or connectivity, you disassemble the part you have and continue to build on the basis on which you started to obtain a workable module. The important part of modular design is that you provide a workable idea that can be put on paper or in a computer model that will benefit the forces in fabrication and installation that is efficient, timely, and cost-effective. These ideas can be developed much more effectively in the design office than to have to rework in the field to accomplish your goal.

CHAPTER 21

Nuclear Design Support Experience

My first experience with design in the nuclear field came as a by-product of other design work. I worked for a company that had a division that fabricated material for the power industry with a specialty of producing large pipe bends. They also had experience in maintenance of power plants, including nuclear sites, for major turnarounds and outages. I was contacted for the design of a pipe trailer that could be used to move small pipe into a paint booth and allow multiple pipe spools to be painted while still on the trailer.

In the past, they had used their own designs and modifications, but the designs had proved to be less than acceptable and were not sufficient to handle the loads imposed by the multiple spools of pipe. Our group was asked to design a pipe trailer that could be used especially for this service, and the design was stamped with a professional engineer's stamp for use in multiple fabrication facilities. After several rounds of design review and modification, we produced a pipe paint trailer that was built per the design and first used in a fabrication facility in Mexico. The trailer was a success, and the design was duplicated for multiple locations including several in the United States.

There was a resurgence of nuclear power plants and facilities in the US, and due to our company's involvement in the maintenance and outage repair of existing nuclear power plants, we were heavily involved in the design and support of new nuclear facilities. A new fabrication facility in Louisiana was constructed to provide modules for the nuclear industry. There were nuclear plants being constructed in Georgia and in South Carolina. I had earlier consulted with the construction group to evaluate the proposed site for the fabrication facility and provided guidance for geotechnical exploration of the site to determine the feasibility of locating the plant at the proposed location.

I also reviewed some of the buildings and structures for the site that were designed by a division of our company. The manager for whom we provided the pipe trailer became involved with the new fabrication facility, which was constructed to provide modules for the nuclear industry. Since we had provided him with a good design previously, he requested support to provide fabrication tools, guides, and supports for the facility. Most of the facility produced structures and modules that were constructed from carbon steel plate, and we provided guides that helped to align and support the plate steel and structural members during the cutting and welding process. A large part of the work was accomplished using laser cutting technics and automatic welding processes and had to be held accurately and repeatedly for the fabrication to proceed at an efficient pace.

A lot of the nuclear design requirements involved containment walls that were primarily made from sandwiched sections of steel plate filled with concrete. These were several feet thick, and the outer steel plates were fabricated in the shop, shipped to the field, assembled, and filled with concrete. These wall sections were fabricated with interior structural standoffs to hold the plates in place. There were a lot of additional measures and details required to produce the wall sections and design checks for the placement of concrete that I will not go into for obvious reasons.

We were also required to provide designs for test panels that could be constructed in the field and disassembled for inspection and analysis to prove the process before it was used in the actual construction. The final wall sections were all welded, but the test sections had to be bolted to provide for disassembly and inspection. Most of the design of the nuclear facility was handled in our North Carolina office. As we provided more and more support to help the overall effort, we were given more and more tasks to accomplish.

Our knowledge and experience with modular design proved invaluable for the nuclear design effort. Almost all design structures were produced in a facility that was nuclear certified and shipped to the various sites as modular units to be installed at the facility site. Each fabrication site had strict rules and guidelines for producing any item that ended up in a nuclear facility. Everything nuclear was regulated by the Nuclear Regulatory Commission (NRC), and there were regular inspections and verifications of the site and the work produced.

Not only was the actual product inspected and evaluated but the amount of supporting documentation and paper trails were enormous. Everything that went into the final product had to be accounted for from the inception to the final product. Every piece of steel plate or structural member was followed from the steel mills that produced the product to the final location and use of the product. Welds were followed from the production of the weld wire and weld rods to the location of the welds on the final product. The materials used in the production of the concrete used on the walls were followed from the material source to the final product and location. Every time anything changed, it had to go through several levels of inspection and approval.

I remember one case involving a weld between plate sections that was inspected in the field and found to be too rough to meet the designated requirements. A request to grind down the weld and repaint was introduced into the repair procedure. It passed through multiple levels of approval, and even I had to sign off as a responsible engineer before the repair could be made. It took almost two months before the repair was approved and completed. If this had been a normal industrial project, the field supervision would have begun the repair and, at most, informed the design office of what was done.

It could have been accomplished in a few hours at most, and the construction effort would have resumed. Incidents like this are part of what run up the costs of nuclear facilities. I am not suggesting we abandon the safety measures required for nuclear facilities, but I think some review of the requirements by knowledgeable persons could reduce the unnecessary cost and time of nuclear facilities. I also realize that those knowledgeable persons are also the same people that benefit from the extra cost, inspection, and delay of the work. After all, most of them work for the government, and I remember the famous line that goes "I am from the government, and I am here to help you." I don't want to sound political because I have been helped and hurt by the government. It is up to everyone to decide for themselves.

Work for this facility producing nuclear support modules led to the request to provide pipe supports for another type of nuclear facility. Because of the START treaty with Russia, the United States committed to the reduction of nuclear weapons. A facility in South Carolina was being constructed for the purpose of converting weapons-grade nuclear material to power-plant grade nuclear material. The design was begun years previously by our office in North Carolina and was partially constructed using design methods for blast walls and so on.

The need for a group to design pipe supports for the facility fit perfectly with our group, and we began as a small nuclear pipe support design group within the overall civil/structural design group. We received loads and layouts from the main design office and performed our own design calculations, reviews, and submittals of design drawings for the miscellaneous pipe supports. The design effort was much more intensified for the supports than it would have been for a regular pipe support. Special attention was paid to the weld size, type, and location. Additional design checks were performed, and typical sections involved a lot of pipe and tubular sections for support. As we provided the design needed for the supports, we were given additional supports. When the main office could not provide manpower for the workload, we hired additional manpower for the Baton Rouge office and continued to provide support as requested.

ENGINEERING

All nuclear projects required extra procedures and requirements, and our design office was no different. Each individual engineer and designer were required to read and follow nuclear guidelines established by the project and the NRC. Training was conducted locally and through online classes. Copies of the required training classes and lists of the personnel completing the training had to be maintained. As classes and training were added to the project, each person working on the project was required to complete the training. This training helped us to realize the need for the procedures we were following and made us more aware of the overall project requirements.

When I retired in 2015, the nuclear design of pipe supports was still continuing, but a short while after that, the government reduced the funding for the project. Our participation as a design group was halted. I do not know the end result of the power plants we supported or the nuclear materials reduction facility.

CHAPTER 22

—w—

Management Experience

My management experience throughout my career has been varied and has been obtained through several different methods. As I look back, I can see that my total experience was a combination of a lot of factors, experiences, and associations that, taken together, formed the basis of how I managed my work assignments and the people who I worked with during the various projects. Even on my first summer job as an engineer, I was sent away from the office to offshore site locations, fabrication facilities, and marine loading docks. While at these locations, I was required to engage the personnel there and coordinate the fabrication, transportation, and installation of the various design packages. I learned to work with people from the suppliers and clients and to manage their expectations as well as the company's expectations for a successful project. I learned to expect issues to arise that required that decisions be made, some that I could make as a representative of the company and others that required me to consult with my management to determine the correct course of action.

I used this knowledge of working with people to help me when I returned to Mississippi State University in the fall and, on a graduate assistantship, began to teach undergraduate lab classes and work

ENGINEERING

for the department professors to grade homework and tests for the undergraduate students. I was assigned to teach an undergraduate hydraulics laboratory class. The class had equipment in the lab that would demonstrate the hydraulic principles of pipe friction flow, laminar and turbulent flow in channels, and hydraulic jump. The students would run a series of tests in the lab and verify the hydraulic principles by feeding the lab results into a computer program for analysis.

At that time, personal computers were not available, and even personal handheld calculators were not sufficiently in widespread use to require the students to use them for classes. The calculations required were studied in the hydraulics lecture class, and the lab was only used to demonstrate the principle. I took it upon myself to provide the basic design calculations, had them put on IBM cards, and placed the cards at the central mainframe computer site for the college. The students then could place a set of IBM cards with the data from the lab on the cards and ask that the data cards be placed with the set of calculation cards kept at the computer mainframe terminal and the analysis could be performed with a printout available to present in the next lab class (see chapter 5).

This ensured that the students all used the same program and we did not waste time having the students write a computer program from scratch and possibly having an error in the calculations that would affect the outcome of the lab results. This is very similar to having a student today analyze data using a known and accepted computer program to arrive at an acceptable analysis. I do not know what was done before I started teaching the class, but for those two years, I used my management skills to improve the analysis procedure.

The use of IBM cards continued for many years, and the computer program submitted as a part of my thesis in 1975 was located on IBM cards. If these lab tests were analyzed today, it would probably be on a personal computer, laptop or iPad, and the data as well as the analysis would probably be on a flash drive or other high-tech memory device.

It may have seemed trivial or minor, but I believe it was the start for me to begin developing management skills. Management is simply the process of dealing with or controlling people or equipment. There are all types of management such as personnel, business, fabrication, construction, engineering, time, schedule, facilities, transportation, training, operations, inventory, inspection, drafting and CADD, safety, sales, etc. I have had experience in all these areas, some more than others, and I will provide some of these as I indicate my management experience.

When I first began my full-time employment with a petro-chemical company, I was part of a team and reported to my management for guidance and assignments. As I gained more experience, I was sent to the field and sent on special assignments where I was expected to manage my work and coordinate with others as needed. See chapters 6 through 12 for my work experiences. I was able to develop management skills related to fabrication of steel structures and buildings. It was my job to provide designs that could be built efficiently and safely. Close coordination and understanding of the process was developed.

I was learning engineering management skills as I worked with teams and as the lead engineer on projects. I began to understand that for a successful project, everyone must do their part and provide input in a timely manner to coordinate with other team members working on the project. I learned the importance of the design feasibility and timing to match the construction effort. Schedules had to be managed to obtain the correct resources and materials and the timely transportation delivery of designs and equipment for the construction of the project to be successful. I experienced inventory management from engineering tools and equipment required to do the job to inventory of construction equipment and supplies needed in the field to complete the project.

I experienced other types of management when I was asked to supervise the drafting and design functions of the engineering department. Inventory of supplies were critical. Management of the equipment was necessary to maintain the flow of work. One of my assignments involved leading inspection teams for buildings requiring major maintenance or rebuilding. Our inspections had to be coordinated with the plant operations since the plant continued to

ENGINEERING

run during our inspections. Close coordination between our inspection crews and the operations team was required on a daily basis

I was the lead civil engineer on an installation of a backup power generator for the headquarters facility of a major corporation, and I was required to coordinate with the state fire marshal office for the permits to construct the generator building and the power bus bars into the main building. Close coordination with the building facilities and the main computer backup personnel for the corporation was required.

Later in my career when I was transferred into construction, the management of over two hundred construction personnel was my overall responsibility. I learned to take steps to simplify the management and streamline the effort to approve the work on a weekly basis. For my management efforts and decisions, see chapter 15. This includes personnel, fabrication, construction, engineering, time, schedule, facilities, transportation, training, operations, inventory, inspection, and safety while in construction. This required a lot of time and effort, but I believe I learned more about management while in construction than anywhere else.

Problems arose every day, and solutions were needed immediately. I had to think on my feet and learned to trust my instincts for solutions. I had a lot of people helping me and giving me advice, but I was responsible for making the decisions and carrying them out. I learned to deal with equipment delays, breakdowns, and shortages. But dealing with personnel was much more difficult.

After I returned from construction, I began to work nights and weekends for a construction company in the New Orleans area designing sheet pile walls as temporary construction while the canals were being reworked with concrete walls and bottoms. This work was totally separate from my day job but did use some of the same design skills. I purchased design software for my home computer and utilized software provided by the US Army Corps of Engineers. I was now coordinating and working with various design firms, the construction firm, and most often, with the Corps of Engineers. I became proficient enough that the Corps of Engineers supervision was sending people to me to guide or explain to them how to submit design projects in the correct way for the approval of the Corps. I worked for this contractor on and off for almost fifteen years, and I continued to develop management skills required to handle the work effectively.

During the period before I transferred to construction and before I started to work for the outside contractor, my wife and I also started a wood-burning stove business. This was in the early 1980s, and we had the outside business for about four years. We set up a full *Inc* corporation (not like the LLCs that you see today). I was the president, and my wife was the secretary-treasurer. I obtained a lot of business management skills and got a lot of things at wholesale cost but did not make a lot of money. It was not the best time to go into business because the interest rates were sky-high. I would borrow money from the bank, buy stoves and accessories, sell them, and then go pay off my ninety-day note at the bank. What I did gain was business management skills and a lot of good knowledge and experience. I learned to deal with the public, hire and support employees, and gained sales experience.

My management skills were moved to another level after I changed companies. I became the lead of the civil design group in the early to mid-2000s and built the design group (both engineers and designers) from approximately ten to twelve people to as much as fifty people at one time. There were still forty-seven people reporting directly to me when I retired in September 2015. Most of the increase of personnel in our group was the result of obtaining new sources of projects from within the overall company. Most of the increase started as a result of one project (see chapter 21). This nuclear design work added a lot of personnel and required additional management skills. Even though I was the direct supervisor of the entire civil group, I utilized the services of various individual people to help me manage the group. All the major projects had engineering and design leads that managed the projects on a day-to-day basis and consulted with me for overall supervision. Within each major project there were personnel that had specific duties related to the project.

I learned that, with more than forty-five people in the group, I had to put in a lot more time to effectively manage the group. My

ENGINEERING

hours increased to around fifty to fifty-five-plus hours each week, and even when I left the office on vacation, I still was contacted very often for one reason or another. I remember one trip my family had taken out of state, and I was called on a Saturday. I resolved the problem, but then I was contacted by someone else on Monday to ask me to provide an answer for the problem I had already solved. I also noticed that I could not solve any problems I was working on unless it was early in the day before six or 6:30 a.m. or late in the day after six or 6:30 p.m. Even though we were mostly on a four-tens schedule, I ended up working usually five-tens or five-twelves and answered problems on the weekends via phone or laptop (they generously allowed me to carry my laptop with me on the weekends). The end result was that I only solved problems for everyone else for the most of the day.

I was not complaining, but it did require good management skills to be able to handle a group this large. I was often asked how I managed to get so much work for the group. We had become the largest design group, even larger than the piping group, which is traditionally the largest. My reply was "It is not always what you do for someone yesterday or last week or last month, but it may be what kind of job you did for them last year or five years ago. Always do your best, and it may result in future business when you least expect it." I also told everyone that I pray a lot.

Good management is something you do continuously for every endeavor, and even though sometimes it may not result in immediate success, in the long run, it will pay off. It is just what you do to handle people, equipment, and especially the situations that you confront every day.

CHAPTER 23

Training Experience

Soon after assuming the supervision of the groups both in design and in construction, I realized that it was my responsibility to make sure that all the people in my group had received proper training to do their jobs. Initially, all the engineers had degrees in engineering and the designers had graduated from technical schools (in some cases, the designers had developed their skills by experience). As codes, regulations, and technical knowledge developed, it was necessary to provide additional training.

Much of this was accomplished by sending personnel to off-site training courses and seminars, and some of it was provided by bringing the courses into the work office for group training. Sometimes an individual was sent remotely to secure the knowledge and materials needed and returned to the office to pass the information on to everyone affected in the group. In the early years, this was usually the training required for an updated revision to the steel, concrete, or building codes and was scheduled for everyone as needed on an annual basis. Some training was needed for different types of technology or methods to be used by the civil group. This included things such as learning to design and specify auger cast piles, the use of direct tensioning devices for steel connections, the use of laser leveling technics, the use of computers for design calculations, and developing skills for advancing technology.

One of the major advancements was the use of computer-aided design and drafting (CADD). See chapter 14 for my experience and training on the CADD system. This training was brought back into the office, and I held multiple classes with the employees who were selected to become operators to bring them up-to-date with the procedures and methods to use the CADD system for producing drawings that could be sent to the field and constructed.

Once the company had switched over to producing drawings electronically that could be printed/plotted on paper similar to the drawings that were created on drafting boards, it changed a lot of everyday functions. The electronic drawings were initially plotted using ink pen plotters and later converted to laser printers/plotters. The pen plotters had to be maintained and serviced, and the day-to-day maintenance was my responsibility. Once we were using the laser plotters, they were serviced by the company providing maintenance for our copiers and reproduction equipment. Previously, if an additional copy of a drawing was needed it had to be copied onto sepia paper. There were machines capable of producing blueline prints from an original paper drawing. These machines could also produce a copy of the drawing that was reproducible if sepia paper was used. Once the drawings were produced electronically, it was easy to provide another copy from the original by simply sending the original print to the network plotter.

When I was transferred to construction, there was another type of training required. The construction forces did not need the skills necessary to produce engineering design and drafting but needed the skills necessary to construct and install the items shown on the drawings. Most of the construction workers had received basic training in their particular area of work, but there was a lot of training required for complying with OSHA requirements and plant site safety rules and procedures.

Whenever there was a large enough construction force dedicated to a plant site, the construction organization usually consisted of a designated trainer, either part- or full-time, assigned to cover industry and OSHA training. However, for small construction groups, the training was the responsibility of the company construction coordinator. While at the South Carolina site, I was the construction coordinator for all capital projects at the site. And at the Illinois site, I was the construction coordinator for a large construction project. At both sites, I covered training for some of the OSHA requirements and all the plant site specific training.

While at the South Carolina site, a safety incident occurred that made me realize that the construction forces did not have access to the plant safety rules. They were not allowed to be issued copies because the safety manual was a controlled document and could not be issued to anyone outside of the designated plant personnel. I felt it was important to include each of the construction personnel with a set of rules and guidelines to work safely in the plant. Using the plant safety rules as a basis and OSHA and federal regulations as a guide, I produced a contractor safety manual of approximately twenty-five chapters that gave guidance for such items as hot-work permits, excavation standards, emergency evacuations, vessel entry, scaffolding, confined space entry, and other plant-specific requirements.

In order to provide the same safety rules as the plant forces were using, I included a copy of the plant safety manual as an appendix to the contractor safety manual and listed it as uncontrolled. Therefore, if an issue came up that would have been addressed in the controlled plant safety manual, I could make available a current copy of that portion of the manual since I was on the distribution list for the plant manual. Each time the plant safety manual was revised, I revised the contractor safety manual and included an updated copy of the plant safety manual in the appendix. Each contractor that worked at the plant site was issued a controlled copy of the safety manual for contractors and was expected to adhere to the requirements and guidelines. This satisfied everyone and kept each construction employee up to date and as safety aware as possible.

While administering annual reviews to the personnel reporting to me, it became obvious that the employees needed certain training for skills that were expected of them but skills they were never trained in. It was assumed that the employees would develop these skills on their own time and effort. At each review period, I evaluated the needs of the employees based on the skills and training they thought they needed the most and began to search for ways to provide the training needed. There were always technical skills that needed to be developed based on the changing codes and regulations and new methods of design, fabrication, and construction. These could usually be addressed by courses offered by the technical societies and organizations. Personnel were provided the opportunity to secure this training as it became available. I had to look deeper for training in the skills needed for people working in groups to improve their effectiveness.

The most pressing skill requiring improvement was the ability to communicate. A large percentage of the requested training from the annual reviews was the need to communicate better. I researched the company's available training related to communications and found one class labeled as *assertive communication*. I was happy with any kind of communication, so I inquired if the class was available and was told that it could be arranged. The instructors would have to be sent to our location from out of state, and there was a substantial cost for the training course to be presented. It would have been an uphill battle to convince management to provide the necessary funds for the course.

I began to search the internet for information on communications, and in a short time, I was able to gather enough general knowledge about communications and assembled a four-hour course on communications. It included the standard items such as the definition of communications, types of communication, barriers to communication, communication aids, etc. I also provided examples and group interaction/discussion to help in the understanding of the course. I presented the course to the entire civil design group, and it was well received. Most employees are very accepting of training that is presented to help them with their jobs if it is shown to have relevance and meaning.

The interaction of the group was extremely important to get everyone involved in the training. After the success of the communications course, I looked at additional training that would be beneficial to the group. I developed a class on teamwork and a class on leadership so that we could derive the maximum benefit from the different design teams and groups working on the various projects.

Some of the technical guidance needed for certain topics was so detailed and involved, a condensed version was needed to be able to grasp the essentials. One of these was the *NFPA 101 Life Safety Code*, which would probably require at least a week to thoroughly cover all the specific requirements. I was very familiar with the *Life Safety Code* and felt that I could extract the specific requirements that applied to our typical design and present it to the group in a condensed form. I was able to do this, and it was presented to the group in the form of a customized class. Other personnel in the company that had a specific need for the classes were also invited to attend.

Over the years I prepared and presented many classes to the civil group and to others in the company on such various topics as checking drawings, checking fabricator drawings, welded connections, bolted connections, CADD design and production, organization of files and resources, procedures for drawing sign out, OSHA excavations, critical lift procedures, and other topics necessary for the functioning of the design group. While in construction, I taught many classes on OSHA training, safety procedures, and classes related to hot-work permits, confined space entry, critical lifts, and plant safety rules and guidelines.

It is very important to provide training and guidance in order to pass on skills and knowledge to others. Sometimes skills are best learned from those who have experienced the situations and can relate to others rather than simply reading a book. Much of my early experience and knowledge was passed on to me from those who possessed the knowledge from their own experience, and it proved to be very valuable in my early years. I will always be grateful to those who helped and guided me in developing my skills and knowledge, and I wanted it to be passed on to others for their benefit.

In the same manner, I hope that this book will help someone else in their quest for knowledge and guide them in gaining experience in whatever career path they choose. Please refer to the Appendix section of the book in which I will include some of my aids and guides for others to review and evaluate and, hopefully, give the reader some helpful guidance.

CHAPTER 24

Proper Engineering Projects

For a successful engineering project, it has to be developed in the correct order with all the parts and pieces intertwined at the proper time while utilizing the appropriate resources necessary for completion. Having said that, it is recognized that a great number of successful projects do not have all the needed elements available at the correct time. Many obstacles have to be overcome to ensure the project will be completed within budget and on schedule. There are unforeseen obstacles that may interfere with a project. Natural disasters such as hurricanes, tornados, floods, earthquakes, fires, and other unexpected occurrences may delay project completions and could cause damage to partially constructed facilities. These problems have to be addressed if and when they occur and cannot always be anticipated.

Some natural problems can be anticipated such as excavations during the rainy season, installation of piling adjacent to river levees during periods of high-water levels, or construction of facilities during winter months in cold climates. These may require additional preparations or extensions of the schedule. Some items can be considered in the project schedule and should be a part of the overall project completion. This means that the schedule for a particular task may need to be lengthened if the schedule allows or additional resources may need to be added to shorten the time required for a certain task. Modular construction may be used in order to begin the assembly of some items in the fabrication shop and shorten the time required in install them in the field.

One of the projects I worked on was a landfill gas recovery project in the state of New York. During the construction of the facility, the project was shut down and the landfill was reopened to receive debris from the Twin Towers disaster in New York on 9-11-2001. This was definitely an unseen circumstance that had to be accounted for in the schedule for the project completion.

The need for a project is usually first developed by the client who is in need of extra production capacity, a decrease in the cost of a product, production of a new product developed from research, modification of a product to satisfy environmental and regulatory requirements, or other reasons justified from the process or research-and-development departments of the client. Sometimes it could be the result of improved methods of production for a well-established product.

The client prepares a request for bid (RFB) package that includes the scope of the work to be performed, the anticipated schedule for the requested bidder, the expected deliverables due to the client from the bidder, and the background documents that define the scope of the work. These background documents will vary greatly depending on the type and stage of the work requested. Background documents (both preliminary and approved) may include but are not limited to plot plans, piping and instrumentation drawings, equipment data sheets, request for equipment quotes, equipment and material specifications, vendor lists, plant surveys, geotechnical reports, existing plant drawings, third-party reports and analysis, and any other documents that help define the scope of work that will aid the bidders in providing a timely and appropriate bid to do the requested work. The request for bids may be very limited such as providing an inspection service for concrete pours or providing a bid to fabricate a specific piece of equipment. The request could also be very large such as the management of a large multidisciplined construction effort that will last for several years.

Some clients may have contracts with various engineering firms to handle projects based on multiyear contracts. The bidding process may only involve the timing and the transfer of information between the client and the contractor. The contractor will reply to ensure that the schedule can be met following the terms of the contract already in force between the client and the contractor. Personnel will be assigned, and dates for transfer of information and deliverables due to the client will be agreed upon. And then the work will begin.

Whenever there is no current contract between the parties, the request for bids will usually be sent to multiple contractors (typically three or four) to select the best bid based on schedule, cost, and quality of service. Most clients do not send out bids to contractors they do not believe will provide acceptable service, but there are exceptions. For example, a contractor that provides quality service may only be able to perform 75 percent of the requested work due to manpower, capabilities, or other reasons but is included in the bid. If the 75 percent of work this contractor can perform is acceptable due to schedule, cost, or quality, they may be awarded the work and a separate contract will be requested for the remaining 25 percent of the work. To compare this with the other contractors bidding the work, all contractors may be asked to provide separate bids for portions of the work. It should also be addressed in the request for bids that any or all portions of the bids could be accepted to make the process fair and aboveboard for all.

Each project is unique and should be addressed individually to determine what constitutes a proper project and how to avoid expected problems and how to address unexpected problems when they occur. I will go through some typical projects to show how the flow of the work should be done ideally and what can be done to make the flow of work smoother and stay on schedule and under budget.

On almost all projects, someone up the line has already determined the desired completion date and cost based on preliminary estimates by the client. This is usually set by the market forces that determine when a product has to be ready for the market and how much it can cost to make sure it is profitable for the client. Despite what you hear on TV ads about how much a company cares about the customer, if it is not profitable, the chances are very limited that it will actually make it to the market.

Almost all the people that have large sums of money to invest in companies and products do so to increase the amount of money they have through profits. There are those people that give away their money through foundations and charities, but I cannot think of anyone who invests in companies or products with the intention of losing money. There are plenty of people that lose money but not intentionally. That is a long way of saying that most projects are limited by schedule and costs before any engineering or construction contractor is invited to participate in the process. Therefore, most contractors try to work within the requested schedule and budget if possible. Often the actual budget is not revealed to the contractor, and the bids are evaluated to determine which contractor is within or closest to the anticipated budget. For those engineering and construction departments within the client company, it may be easier to stay within the schedule and budget since the departments are all working for the same company.

I will point out that even though there is an effort to accurately predict the cost and required market date for a product, the time between the product inception and the actual ready-for-market date may change the profitability of the project. I have worked on projects that could not be completed fast enough, regardless of the cost, because the potential profit was enormous. Incentives to vendors, additional manpower resources, overtime for all parties involved, and any other means to bring the product to market quicker were employed. The opposite was also true. I have worked on projects that were profitable when initiated but not profitable when ready for the market.

I spent a good deal of my career working in the petro-chemical industry, and success is often determined by the timing of a project. A lot of products are successful, and their life span goes on for many years, especially with licensed technology and patents on the products and processes. These successful products allow the companies to invest in other products that have a high rate of return but also have a significant potential for failure if the timing is not correct.

I worked on a project once that had a successful design and construction phase with very limited problems. The product was installed in a new facility in an existing plant site. A new warehouse was constructed to hold the product while waiting for shipment to the customer. Between the time of the start of the project until the plant start-up, the economics changed and the product could not be sold for more than it cost to manufacture. After the plant start-up, the warehouse was filled and the plant was shut down, never to run again. A few years later, the facility was modified to produce another product.

Because of the rapid pace of the execution of projects, it became important to be able to make good decisions at a fast pace that were conservative enough to withstand minor adjustments if needed. In the petro-chemical business, most projects will only last for five to ten years before they are upgraded, modified, or completely reworked. Engineers working on other types of industry projects often take more time and effort to complete their work because of the schedules and the longevity of the completed work.

To have a successful project, it is important to receive and complete tasks in an orderly manner. Here is an example of a typical engineering project that will demonstrate how the parts and pieces are fit together to complete the work. The engineering group receives a request for bids for a project. A project manager is assigned to oversee the project for the engineering department, and the engineering manager assigns the design work to the various disciplines. These design disciplines usually include process, piping, mechanical, civil, electrical, and instrumentation.

If the project is large enough, a dedicated project team is assigned, and those people on the team will work on that project only. If it is a smaller project, the project manager may assign a project engineer to work on the project and the various design discipline supervisors may assign a particular design lead for the project. For this example, we will assume the project is sufficiently large enough to have a dedicated team. Other disciplines such as cost control, scheduling, estimating, information technology, etc. will also have personnel selected to support the project.

Team meetings are scheduled, usually each week, at a specified time and place. For those who cannot attend, there may be a phone call-in set up so people from remote locations can participate in the meetings. This will include anyone on a trip or anyone that will participate only as needed such as a construction representative or a representative from the client. For major projects, the clients are a part of the regular meeting each week. The meeting is conducted by the engineering project manager or their designee. Meeting notifications are sent to each participating party with an agenda of the topics to be covered in the meeting. If the designated member cannot attend, a representative is usually sent to participate.

A typical agenda will list the project, the meeting date and time, and the attendees. It will be divided into current work completed during the past week, work planned for completion in the future (usually called a two-week look ahead), and issues of concern to the individual group or to the team as a whole. The agenda sent out for the meeting will only have headings for these topics and will be filled in after the meeting with current information. This also serves as a reminder to each member that they should be prepared to report on work completed and work planned and to write down any concerns they have. By preparing each section of the agenda prior to the meeting and attending the meeting with the items written down, it makes for a much smoother and efficient meeting. A copy of the agenda items discussed by each representative should be turned over to the project manager at the beginning of each meeting.

Most meetings are started with a safety topic. This should be no more than five minutes in length (one to two minutes would be better if practical) and can relate to the office, the field, or the current project as appropriate. Each representative member of the team will have a brief period to discuss current and future work and their concerns with the team. Responses from other team members should be brief and to the point. The team meeting is not a place to solve problems but only to share work progress and concerns. Responses from

ENGINEERING

other team members may agree or disagree and should be taken out of the meeting for lengthy discussions. Problems can be addressed separately by the specific team members that are involved, and any decisions made should be reported by phone calls/e-mails and can be reported in the next team meeting.

There needs to be a mechanism for solving issues brought up in the meetings. An action list is included as a part of the agenda that lists the issue, the date the issue was presented in the meeting, the assigned personnel or group to resolve the issue, and the date the issue was resolved. This might include topics such as "issue approved P&IDs," "develop preliminary cost estimate," "issue RFB for geotechnical report," "develop piping specifications," "determine feasibility study of prefabricated motor control center," "determine if foundation package can be issued two weeks early," "revise schedule due to late receipt of reactor drawings," etc.

Minor problems that do not affect the cost or schedule should not be included on the list and can simply be reported in the meeting. The main purpose of the meeting should be communication. Let everyone know what you are doing, what you plan to be doing, and what concerns you have that may prevent you from completing your portion of the work. It is the responsibility of the project manager to keep the meeting on track and running efficiently. If each member of the meeting is prepared and follows the agenda, most all meetings can be handled in one hour or less. Keeping meetings short and to the point will allow more time for members to get back to work on the items for which they are responsible. After the meeting is over, the project manager will assemble all information provided by the team members along with any new information or details from the meeting and issue a final meeting report to all participants and interested parties.

There are exceptions to the "one hour or less" guideline. At times there will need to be lengthy discussions about portions of the design, and these can be accomplished at a weekly meeting. Such things as reviewing the model, preparing the schedule, reviewing the schedule, and discussing third-party design packages can be included in an extended meeting. Additional personnel may be required to attend and will include engineers, designers, support staff, and any third-party representatives as needed.

Since I am most familiar with civil and structural design, I will focus on the steps taken by the civil design group during a project. The civil and structural group will be concerned with site layout and drainage, structures, buildings, foundations, piping and equipment supports, and plant roads and railroads. Other disciplines will have similar tasks but will be tuned to their specific requirements. The process group will be concerned with specifications for piping, instrumentation, equipment, and development and adherence to the P&IDs.

The piping group will be working on piping specifications and layout, will work with the process group on P&IDs, and will be concerned with pipe stress analysis. The mechanical group will focus on equipment specifications, the P&IDs, and development of preliminary equipment drawings required for bids. The electrical group will be concerned with electrical materials, equipment drawings and specifications, grounding, and lighting and power requirements. The instrument group will focus on the P&IDs and instrument specifications and will work with the piping group for in-line instruments.

The estimate group will prepare and update the project estimate as needed. The cost control group will monitor cost and keep all disciplines advised of current activities and cost trends. The project group will manage all phases of the project and coordinate activities between the groups. The project manager or their representative should be aware of all interaction between the engineering department and the project client. Some communication will be directly between the discipline groups of the department and the client representatives, if it is more effective, but the project manager should be aware of all communication.

In most cases a perspective client has issued a request for bids (RFB) document, and the information is passed on to the design leads and support staff for their input. Estimates for each discipline are provided as to the quantity of materials, the required equipment, and the time to complete the design along with the total manpower required from each group. These inputs are used to provide a preliminary project estimate and cost for the work to be completed and a scheduled completion date. This estimate is reviewed by all groups and agreed upon. Management will review the estimate to evaluate the manpower available and the required manpower to perform the work.

Management will also determine how well this project will fit with other ongoing work and what appropriate multiplier will be used to determine the final cost and schedule to be presented to the client. This multiplier is adjusted to reflect the work on hand or lack of work at the present and the relationship with the prospective client (i.e., Is this an ongoing client relationship? Are we trying to win a new client? Do we have more work than we can handle at the present, and will it require hiring additional people or work a lot of overtime to complete?). All these factors are considered, and then management selects an appropriate multiplier and submits the bid. If our company is the successful bidder, plans are made to prepare for the upcoming project.

In the civil/structural group, the group leader is made aware of upcoming projects. For dedicated projects, a civil design engineering lead is designated, and all work will be coordinated through the engineering lead. For larger projects, a civil designer lead may also be designated to coordinate all designer functions for the civil group. For a typical job, the engineering and designer leads are provided with the project information and requested to attend a kickoff meeting. The meeting is facilitated by the project manager, and all engineering and designer leads, support staff, and project management are present. For a dedicated project, both engineering and designer leads are represented. Information is distributed to the attendees and may include the following documents:

1. The scope of the work to be performed
2. Requested schedule of engineering department work
3. Deliverables to be provided by the engineering group to the client
4. Plot plan of the work to be done

5. Existing site drawings of the area to include buildings, structures, roads, pipe racks, utilities, and other drawings required for the completion of the requested work
6. Existing equipment drawings associated with the work
7. Existing P&IDs for the work area
8. Existing piping, mechanical, electrical, instrumentation, civil and structural drawings for the work areas
9. Any plant surveys available that indicate plant stormwater drainage and underground pipes and electrical duct bank location
10. Any existing geotechnical reports for the work area
11. List of all key personnel for the engineering department, construction supervision (if appropriate), and the client
12. List of all dedicated and approved vendors by the client
13. Any requirements specific to the client and this project
14. List of any additional third-party companies that are providing services to the client for this project
15. Future scheduled meeting time and place

After the kickoff meeting, the civil leads will get together with the engineers and designers and discuss the upcoming work and assignments.

For large projects, areas will be assigned to designers and engineers. The most important thing at this point is to obtain the information needed for the design. If it is a rework of an existing area, the information for the existing foundations, drainage, structures, and layout of the site is needed to plan the additional modifications. If it is essential, new site additional information will be required such as soils information and proposed layout of the project. In all cases, the equipment information is required in order to support and provide protection and maintenance of the facility.

At the early stages of the project, there may be only preliminary equipment information, and it is used and verified at a later date when the equipment approval drawings are received. Equipment may be supported at grade by foundations or in a structure (usually with steel supports) as presented in the proposed layout and arrangement for the project. For equipment and tanks supported at grade, there may be a diked wall surrounding the tanks to contain any potential spills. For equipment located within a diked area, called a tank farm, there will need to be foundations for tanks, pumps, and other related equipment.

To provide for the piping of liquid material to and from the tanks, there will need to be pipe supports, both structural steel, and foundations within the tank farm. Depending on the size of the project, a series of multiple pipe supports, called pipe racks, will connect the different areas of the facility. These pipe racks will contain bracing for wind and temperature loads and will connect the production, utility, tank farms, load racks, storage areas, and other separate areas of the project.

The civil/structural engineer has to design their portion of the facility backward in most cases. The engineer receives equipment and pipe sizes and loads in order to design the structural supports and pipe racks before they are able to design the foundations required to support the structures. The engineer must know the sizes, weights, and capacities of the tanks in order to design the foundations and determine the sizes and heights of the diked walls to contain spills.

The first thing the construction contractor needs is the design of the foundations and the underground drainage. Both the civil/structural engineer and the electrical engineer (for underground electrical duct banks) must coordinate the location of underground items before the foundations can be issued for construction. On most projects with a fast-paced schedule, it means the civil/structural engineer must design the foundations based on preliminary information and verify the information before releasing the drawings for construction.

If there is a change in the equipment loads or location, the foundations might have to be revised in the field. For this reason, the engineer usually has a little bit of extra built into the foundation, but it does not always work out that way. Most projects that have a tight schedule are willing to make some changes after drawings are issued if it means getting the entire project completed at a faster pace. The foundations, structures, and any other part of the project is always made safe and serviceable before the project is completed.

For a typical project that has an open structure to support equipment, the proposed structural steel layout is begun in CADD with the designer and engineer working together to develop the design. The designer is primarily responsible for the drawing layout and, depending on the level of experience, can develop the frame of the structure with columns, girders, floor beams, and support beams for the equipment. The designer will first work in the CADD model only so it can be checked against the other discipline designs.

The engineer will usually develop a model of the same structure and place in the computer analysis program for solution of the structure. If any of the structural members have to be revised to meet the requirements of the analysis, the members are changed in both the CADD model and in the analysis model until they match. The same procedure is followed for the foundations and any other items. As equipment information is received, the models are updated as needed.

At this time the designer and engineer will develop a bracing scheme for the structure and include it in both CADD and analysis models. Depending on the equipment and associated piping, the bracing for the structure may utilize a moment frame, a full-bay braced frame, a knee-braced frame, or a combination of any of the three bracing systems. The bracing system chosen will also have to consider the location of any pipe runs (called pipe chases), cable tray runs, major process piping, and maintenance of the equipment.

If the structure has an access platform above grade at one or more levels, the access stairs and ladders will have to be considered in the layout of the structure. It is very important to locate the access for stairs and ladders early in the project before the space is taken up by equipment and piping that cannot be easily moved at a later date. All these layouts should be loaded into a master file containing all the disciplines' work to date to continuously monitor for possible interferences that can be resolved in a timely basis. It is best to load each individual CADD file into the master file for the area at least once a day.

Once the equipment information is approved and the layout is satisfactory to each of the design disciplines, the engineer should

ENGINEERING

proceed to developing a final design for the structures, foundations, drainage systems, roads, pipe racks, buildings, etc. Each area should have a separate design but can be combined with similar designs, such as a design for all pipe racks or a design for all API-type tank foundations. The design calculations should include the scope of the work, the design criteria, the equipment information and loading, the method used for the design, the design input and output from computer software programs, any hand calculations performed, and the conclusions from the design calculations. The calculations may also include copies of the plot plan, the equipment layout, and copies of the geotechnical report, and any other third-party report pertinent to the design.

These calculations should be in a standard form as accepted by the company and should be provided to a qualified engineer who did not provide the calculations for this particular design for review and checking. The calculation cover sheet should indicate the project, the title and number of the calculation, and the revision number and include a place for the date and signature of the preparer and checker of the document.

If the design calculation is revised, it should indicate the area of revision and should have the revision number and the date of signature of the revising preparer and checker. This calculation document should be a part of the deliverables to the client for the project. For any given project, there will be multiple calculations for all items designed by the engineer.

After the CADD model has been approved internally, the designer begins the process of creating drawings from the model, placing them on standard-size sheets, and annotating the drawings with notes, dimensions, and other text relative to the drawing. At this point, the drawings can begin to be reviewed by the engineer and other design personnel in all disciplines. The civil/structural engineer and designers will also receive drawings from other disciplines for information and review.

Once the designer has completed the work on the drawings and is in agreement with the engineer, the drawings are submitted to a drawing checker. The checker also receives the documents used to prepare the drawings such as equipment information, engineering sketches, drawings from other disciplines that affect the design, etc. The drawings are checked for accuracy, completeness, and adherence to specifications, design criteria, and drafting standards. The designer also will prepare multiple standard drawings such as typical steel connection details, concrete paving details, and pump foundation schedules that should be checked along with the other drawings. Any conflicts with the drawing checker or the calculation checker should be resolved. In many projects, the drawings are issued for review or preliminary for information prior to completion. and these drawings should always be stamped "Not for Construction." Once both drawings and calculations have been checked and approved and all comments are resolved, the drawings are ready and should be stamped "For Construction" with the date and issued.

This procedure is followed for each area of the project. Weekly meetings are held to update the design progress, and the information received is continuously used to update the design calculations and the drawings. Depending on the project schedule, the design construction packages may be issued with a single discipline package such as "Pile Construction Package." Or it might be multiple packages such as "Underground Construction Package," which could include piles, mat foundations, spread footings, underground electrical duct bank, electrical grounding, and underground natural gas supply including multiple disciplines. Each company will have their specific procedure and guidelines for issuing construction packages and drawings, and these should be followed. Sometimes the client has additional or different procedures, and these must be utilized for each individual case.

In order for the project design to move smoothly forward, there is a desired sequence of events and information received. At the beginning of the project, the scope and a firm schedule must be established. A proposed plot plan, equipment arrangement, and the facility layout must be accepted. As the design progresses, these items may change as needed, but here has to be a starting point that all parties can agree on to get things started. There should also be an equipment list that details the capacity, weight, and overall dimen-

ENGINEERING

sions of each item required for the project. All items will not be fully defined, but there has to be a preliminary estimate to begin design.

The design team will take the information available and begin to layout the structural supports and foundations. It is at this point that equipment supports can be developed that will support all pieces of equipment in a combined structure. It can also be determined if the structure or individual pieces of equipment will be best supported by individual foundations, combined foundations, or mat foundations. Modifications to the design will be required if the equipment information is late or if it has changed to significantly alter the design. Input from other disciplines that require extra space or additional supports may change the design. Client decisions often alter the required supports and may increase the cost or schedule of the project. Each of these items have to be addressed in an attempt to stay within budget and on schedule. A good schedule will allow for these changes by adding extra resources or working overtime to keep on track.

After the initial design is complete and has been reviewed internally for both engineering design and CADD drawings, it is usually sent to the client for review and comment. Once the client has provided comments on the design and drawings, the design calculations and drawings are modified as necessary. If the comments cannot be incorporated, they are discussed with the client, and then the final issue of the calculations and the drawings can be provided. This may occur multiple times if the project is issued in several packages to support the early construction of certain parts of the project.

Civil/structural packages are usually issued first along with electrical and other underground installations. There may also be early issues of prefabricated packages such as modular control rooms, self-contained treatment packages, and other specialized units that can be fabricated off-site and delivered for installation at a later date. Once the civil/structural design for foundations, structures, roads, railroads, and pipe racks are issued, the equipment, piping, electrical, and instrumentation design packages will follow as needed for the field installation.

For larger projects, there is usually a design review that involves the project team, design teams (engineers and designers), supporting disciplines, construction representatives, and the client representatives. The CADD design model is reviewed on a presentation screen, and each discipline design is reviewed for completeness, access, functionality, and the relationship to other discipline needs. All representatives including the client are presented with the design package and given the opportunity to ask questions, provide comments, and approve the design. Any comments or requests for modifications are listed, and each comment is addressed before the final issue of the drawings. Smaller projects may only need one design review, but larger project may need multiple reviews at selected stages of the design to provide comments and direction as the design progresses.

Once the construction of the project has begun in the field, problems may occur that need attention. Any time there is excavation and construction below ground, there is the possibility of interference. If it is in an existing plant site, there may be underground pipes, electrical duct banks, buried debris from previous sites, and natural obstacles such as buried logs, a high water table, weak soil layers, etc. If any of these are found, it could result in the modification of the design. If the design is based on dimensions and elevations from previous drawings provided by the client that are incorrect, the design may have to change. If the equipment as approved and purchased varies greatly from the equipment information provided at the time of the design, it could affect the design.

Sometimes in the rush to complete a fast-tracked project, the design team gets ahead of the information. I have worked on projects that were in construction and the process design was not fully complete. We were building the facility and hoping that the final process the design was based upon could be finalized before the construction was finished in order to make the last few adjustments workable. I have worked on a plant that required additional land to be purchased because the size grew during the design, and the facility being designed (soon to be constructed) would not fit within the current property limits. I have worked on the design of a facility that was requested of our design team after the construction of the site had begun in the field. The construction forces had begun to backfill and level the site before the foundations were designed. I have worked on projects that at the request of the client, the facility be placed within the confines of a certain area that was not sufficient to contain the facility.

Not every job has these types of issues, but there are usually some surprises and unknowns that will affect the cost and schedule for most projects. They have to be handled when they occur. If all unexpected occurrences could be foreseen, they could be planned for and accounted for in the cost and schedule of a project. Then all projects could be handled by computer programs and guidelines, but that is not the case. If all these occurrences were designed for regardless of the likelihood of happening, the cost and schedule for the projects would be enormous. Even when all expected and possible events are considered, there is still a chance of a "perfect storm" event that could cause problems when several multiple events occur simultaneously. That is the reason there will always be the need for human engineers to solve the problems no one expects to happen.

On some projects there may be a need to provide a set of "as-built" drawings to reflect the actual design and dimensions installed in the field for use by the next project in the area. These changes could be the result of modifications due to field interferences, equipment changes, client request, or other issues that change the drawings. If these changes affect the design calculations, they should also be modified. After the construction is complete and the project has successfully gone through start-up and is producing product, the documentation for the design should be passed on to the client as a part of the deliverables for the project. This usually includes the specifications, standards, design calculations, "issued for construction" drawings (CADD model as a minimum), and any other documents requested by the client as a condition of the project award.

As members of the design teams gain more experience and become more familiar with repeat clients, the incorporation of possible expected events during the design phase of the projects will help to reduce the unexpected surprises and will provide more accurate cost and schedule for the project. There are companies that consistently bid on projects and only bid specifically on what is requested. Their philosophy is based on the assumption that if they can provide a low bid, the additional events or requests from the client can be made up utilizing change orders. I tried to think of the actual things that would occur and base any of my estimates for time and manpower on what the actual cost and time required would be.

I believed it was up to someone else on the project management to decide how to bid the project. If I told them the actual cost and time required, they could determine if they wanted to bid the work close and have a better chance of winning the contract and use change orders if they were needed. By presenting management with the estimated actual cost and time, I was alerting them to what could be the outcome. Even when all things are considered, there are usually things that will require change orders for lump sum bids.

A *lump sum bid* is a bid that the contractor agrees to perform the work stated for a fixed price. Any additional work or modifications to the work will require a "change order" approved by the client for work that will affect the cost and or schedule of the project. The other typical type bid is a time-and-materials bid or cost-plus bid in which the contractor agrees to perform the work stated based on an estimated time required and amount of material required. If the time or material required is different from the amount estimated, the contractor is paid at a specified rate and the cost of the material plus a specified percentage over the material cost. These types of bids usually have a set limit not to exceed a certain amount.

After the project is completed and the deliverables are provided to the client, the design files are archived for future reference and the temporary files are stripped from the CADD and engineer files to make room for the next project. Often a scheduled meeting is held with the project team, design team, supporting disciplines, and the design management to discuss the previous project. Items discussed are what went right and how we can continue these practices on future projects, and what went wrong and how we can prevent these things from happening in the future. These are not meetings to find fault but to improve on weaknesses we may have and reinforce our good practices.

CHAPTER 25

Thoughts on the Engineering Profession

I have heard it said that "Engineering is an honorable profession," and I believe that to be true. In order to become a professional engineer, there is a requirement for intense study and preparation and a dedication to serving others to benefit the public. Most engineers have a desire to create, develop, and provide solutions to problems that will further society. This is true if they are in the fields of civil, environmental, mechanical, instrumentation, chemical, electrical, aerospace, computers, or other related fields.

Since I am most familiar with civil and structural engineering, most of my thoughts will center around these disciplines. The benefits to the public are most evident when addressing safety, and most everything a civil/structural engineer does is related to safety of the public. This ranges from providing foundations that are adequate to support loads and prevent failures of structures and equipment to providing safe structures and buildings that will withstand anticipated loads from wind, flood, snow, ice, fire, explosions, earthquakes, and other natural disasters. Adequate drainage systems are required to prevent flooding, and containment systems are required to prevent spills that would harm people and the environment. Safe highways and railroad and mass transit systems are needed to safely move people and materials from place to place during normal daily activities. Unless you are an astronaut on a training mission or temporarily in space, you are subject to the forces of gravity and must be supported in some fashion from the earth or water below. And that introduces the need for a civil/structural engineer to provide the necessary support.

Everything we do today is affected by the work of an engineer. If you went to work at your job today and used a vehicle or some form of transportation, an engineer was responsible for proving the means of transportation. Your car was developed by a team of engineers including structural, mechanical, electrical, industrial, and chemical engineers to provide the final product. The road you drove on was designed by a civil engineer. The traffic lights to safely guide you to your destination was the product of a highway civil engineer and an electrical engineer. The parking lot or parking garage used while at work was designed by a civil/structural engineer. The office or plant site at which you are working was a combination of multiple engineering efforts.

If you work from home, you are surrounded by the efforts of civil, mechanical, electrical, and chemical engineers. If not for the contributions of engineers, we might be riding our horses to work, reading at night under an oil lamp, and doing everything manually without the aid of machinery. Do not misunderstand me. A lot of inventions and ideas have come to light by those other than engineers. In order to develop the ideas to their full potential safely and economically, the role of the engineer cannot be overlooked.

A lot of students when first enrolled in the engineering courses do not succeed but often transfer out to other fields of study. This is no reflection on those who select other areas for a career path but rather a fine-tuning of the requirements versus the commitment to a particular career goal. There are also many people who, after pursuing other career paths, find that they really want to become an engineer and transfer or enroll in an engineering field of study. This is especially true of those who began their work in construction and

ENGINEERING

realized the desire to develop projects rather than just install them. This can also be the case for those who have prior military experience and want to get a formal education to learn more about the skills or experience that they have developed while in the service. Military service or work in fields such as construction will also help develop discipline, and this is a much needed quality if pursuing a career in engineering.

The engineer is often the watchdog for companies that produce products or services for the public consumption. They make sure the products and services are functional, accurate, and appropriate for the customer. If the company management wants to take shortcuts that might provide defective products to increase the bottom line, the engineer can make sure the products are adequate for the intended purpose, correctly produced, and can be used safely by the public. Good conscientious engineers will always protect the public.

As in any profession, there are those who do not uphold the required standards of the profession but will be corrected and weeded out by the professional societies that maintain the high standards in the individual engineering disciplines. In order to provide engineering services, a professional engineer must be qualified, secure the appropriate training and education, have references from their peers, demonstrate engineering skills while under the supervision of a professional engineer, and be approved by the engineering board in their respective states of practice.

An engineer typically studies a rigorous set of technical courses in order to understand the physical, chemical, and biological forces that must be utilized in order to produce products and services for others. In the field of civil engineering, the physical forces that act upon foundations, structures, and drainage systems must be considered. Courses deal with forces, the interactions of the forces, and the resulting reactions and capacities of the materials used in the foundations, structures, and drainage systems. The college technical courses help prepare the aspiring engineer and provide for the background knowledge and techniques that will be required for analysis and design.

Once the engineer has mastered the basics, they begin a career of learning what engineering is all about. Each example, analysis, and design performed under the guidance of an experienced engineer will help develop the skills needed to become a professional engineer. The engineer does not stop learning once they have performed a task a few times but will continue to learn throughout their career. The best mix for an engineering group is a combination of young, recently graduated engineers and a group of experienced engineers with at least five to ten years of experience, preferably twenty years or more. The young engineers are familiar with the latest techniques and are usually faster at solving the problems when presented with the latest software analysis programs. The experienced engineers can help direct the team toward the best method of solution and design based on their years of experience and knowledge.

There is also a lot of pride in the work performed by engineers. I can ride down the major streets and interstate highway systems of nearby cities and view installations in which I had a part in the design and construction. I can also look at Google maps worldwide and locate my designs and installations. It gives me a lot of pride in the knowledge that I used my knowledge, experience, and engineering skills to design and construct something that benefited mankind. Even though my name is not physically on the installations, I still can proudly know that I made a difference. I can also take pride in the many people that I have been associated with and provided guidance to over the years and how their accomplishments will continue to grow as they continue in their chosen field of professional engineering, design, and construction.

CHAPTER 26

Postretirement Experience

I retired after forty-two years with the same people I had started to work with after I received my master's degree. I was looking forward to retirement where I could enjoy the things I had put off because my work schedule did not allow time to devote to other activities. I had a shop built in my backyard and was getting it set up when a flood set me back. I was just getting things back in shape. My time could now be spent on finishing the shop and stocking it with woodworking equipment.

I was enjoying my leisure time when a friend called me and asked if I would talk to someone about a job with another company. I was not looking for a job but thought it would be interesting to talk with them to see what they were seeking. Maybe there was someone I knew of that might be a match for their requirements. The company was smaller than the one I had left but essentially doing design for petro-chemical type projects with a specialty in plastics. I met with the president and the engineering manager who told me what they needed. I relayed to them that I had just left a company with forty-seven people working directly for me and that I did not want to supervise a group. I also stated that I had been working fifty to fifty-five hours per week and I did not wish to work a lot of hours.

In short, if I were to consider a job with them, it would have to be on a more relaxed schedule and pace. I also felt that I was fully capable of handling anything they could give me to do. To my surprise, they offered a job with no one reporting to me based on fifteen to twenty-five hours a week averaging around twenty hours per week and no specific responsibility. I was being asked to check engineering design and drawings and be a mentor to younger engineers and designers as needed on a flexible schedule. Since they had met all my requests, there was nothing left to do but begin work with the new company on a part-time basis.

After the paperwork was completed, I started a few weeks later at the home office located in Baton Rouge. I did not actually meet the civil department manager for a couple of weeks since his office was located in Texas. There were engineers and designers at the Texas location, and my immediate supervisor would visit every couple of weeks. We would talk as needed over the phone.

I did not need a lot of supervision since I was more knowledgeable and experienced in civil/structural design than anyone else in the company. I got along very well with my supervisor and often discussed projects and workload after hours at home on the phone with him. Our parent engineering company was based in Japan, and we often designed and reviewed projects for the parent company with most projects being constructed in either Louisiana or Texas. I had PE licenses in both states and was a valuable resource for the company to PE stamp drawings and make sure the designs were compatible and understandable to construction firms working in the US.

Since a lot of the work my new company did was related to plastic production design, I learned a new type of industry that also had its own type of issues and ways to properly design and support the equipment, steel, and concrete to which I was familiar. The technology for these plastic plants was licensed worldwide, and our company was an experienced design firm that knew how to handle these projects and, therefore, bid on and was awarded a lot of these design projects worldwide. As a result of the projects being either partially designed or fabricated overseas, we received a lot of requests to review and stamp drawings. This included a lot of miscellaneous supports and platforms, turnkey equipment packages, and foundation designs for equipment with metric loads and dimensions.

Most of the plastic plants involved dynamic forces that resulted from the production of the various forms of plastic that were mostly inert and safe in the final forms. A full understanding of the forces and reactions involved were essential to be able to develop a proper support structure for these plants. It was another aspect of my engineering knowledge that I gained late in my engineering career but one that was exciting and challenging.

When I started with the new company, I expected to work for one to two years and supplement my income to be able to provide equipment for my new shop. I found the job to be fun and I enjoyed the work. My work gradually moved from checking the drawings to reviewing and checking the work to include PE stamping the drawings. I was also utilized to perform specific design tasks when no one else was available or when no one else had the capability to do so. I was also asked to help develop project scopes, review proposals, and do detailed design estimates for projects the company was asked to bid.

Even though I was not seeking additional work, it seemed to gravitate toward me, and I was happy to support the company in any way I could. I was also utilized as a company representative when dealing with permits, governmental agencies, and outside contractors and fabricators. I was assigned to work as a consultant for an outside company that required my specific experience related to foundations and soils.

In order to mentor others, training classes were conducted for the group on such topics as weld design, life safety code, constructability, AISC guidelines, etc. I was also able to present classes to the entire company on such topics as checking drawings that were applicable to everyone in the design groups that produced construction drawings. One of the requests from the engineering manager was for each discipline to produce a set of design guidelines to be used consistently by each group for all projects.

Once the design guides were produced and accepted by the department heads, the guides would be available on an internal computer drive for use by all the group members and available for viewing by others in the company. Once approved and placed on the internal drive, the guides could be modified but only with the approval of the department head and the engineering manager. When the workload was slow, I asked my supervisor if I could start working on the civil design guides and was given permission to begin work as time permitted. This was rather easy for me since I had developed similar guides for work at previous companies and most of it was in my head already. I began gathering information and exchanged a lot of the early drafts with my supervisor and gained preliminary approval.

I was then told that the guides had to be in a similar form of the other groups for approval. I reviewed one group's guides (about a half dozen or so) and consulted with an intern who had been given the responsibility of gathering all the guides together for submittal to the engineering manager for approval. Since these guides had been approved, all subsequent guides had to match as far as format and form. I rewrote and arranged the guides to match the approved guides. I began the process of placing the guides on a common drive so everyone in both offices in Texas and Louisiana could review, and we discussed them at weekly meetings. Comments were added, and agreed-upon drafts of the guides were placed on the common drive for access by all.

Some of the guides included attached forms and flowcharts to further enhance and aid in the understanding of the guide. A critical form was to be used at the group design kickoff meeting that listed the scope, proposed budget, and project team members. A flowchart was provided that indicated the flow of the design work from conception all the way through design completion and archiving with references to key design guides and the order to which the design guides were to be utilized. Even though there were comments, additions, and corrections from everyone in the civil design group, most of the effort in producing and organizing the guides was left to me.

After three-and-one-half years, the work slowed down and there was not enough work to keep me fully occupied. I could have stayed on, but the work I was hired to do was being done by another person. I was not able to hold training classes as much as I would have desired. I finished all the projects on which I was working, turned all the design guides over and placed them on the common drive, and cleaned out all my files (including files left over from the previous employee that sat in my cubicle). I also cleared all my files and forms with HR. I did not close the door on my employment with the company but I made a break so that if I never showed up again, there were no loose ends.

I was back to retirement and working on my shop and other things that I had put off while working. I was called back by the previous company and asked if I could return for a special project for which I was needed. At the time I had just undergone rotator cuff surgery on my right shoulder and had my arm in a sling. The doctor had not released me to drive, and I was unable to return. About a month later when I no longer had my arm in a sling and was able to drive and had use of my shoulder and arm, I called the company back and asked if they still needed me. The answer was yes, and I agreed to return—not as an employee but as an outside contractor with no benefits that would be paid by the hour based on an approved time log with no withdrawals for taxes, social security, etc. All that would be on me to pay separate.

The project involved the rework of structures and supports within a plant site due to a disruption of production and a modification of design loads and forces. This particular plant was shut down, and the client desired to return to production as soon as possible. The project involved the assessment of the existing supports and structures with modified forces and loads to determine if the existing structures and foundations were adequate or if they needed to be modified. Some of the modifications required extensive design changes and were developed as needed.

I was extensively used to review and PE stamp drawings for the construction forces in the field. A portion of the project involved reviewing engineering calculations and drawings from an overseas design firm that needed someone in the local area to provide oversight, review, and PE stamping for their portion of the work. I became the responsible engineer for this overseas firm. As the project progressed, I found myself assigned to various tasks because of my experience or capability, and I worked on whatever I was available to do.

When the request for my services was made initially, I assumed it would be for a two- to three-month duration and certainly not over four months. Most of the work was slowing down, and I felt that there were other people that could handle the work I was doing. After working on this project for over seven months, I let management know I was returning once again to retirement, probably for the final time. I still am registered in three states with my PE licenses, and I could use them on a one-time basis if the need arises. But I am content to stay at home, work in my shop, and talk engineering with anyone that wants to discuss the subject.

CHAPTER 27

A Look Back on My Career

When I first decided to become an engineer, it was for two reasons. The first reason was to provide a source of income for myself and my future family. The second reason was my deep desire to solve problems and design solutions for the challenges that were presented to me. I had always enjoyed building things from my childhood to the present day. Not all my solutions were the best but I built on each attempt with a better result and continued to learn as I progressed.

On the farm, I built sheds and helped construct buildings to house farm animals and to provide for storage of corn and hay. When I was in college, I helped develop lab instruction techniques and learned to construct items out of sheet metal, copper pipe, and lumber while working in a maintenance shop. After becoming a graduate engineer, I developed skills in analyzing structures and providing designs for earthwork, foundations, piles, concrete, steel, wood, drainage systems, pre-engineered systems, roads, railroads, retaining wall systems, and composite designs of many integrated systems. I learned to design for wind, seismic, temperature, and a large variety of soil conditions and requirements.

I developed a knowledge of building codes, steel and concrete manuals, architectural details, and modular design. I became familiar with safety codes and ADA requirements. I developed specific skills to design for load cells, nuclear level devices, lifting lugs, and critical lifts. I learned the use of precast and prefabricated components and when to use them properly and efficiently. I developed the skill of checking steel shop drawings, concrete rebar schedules, and equipment drawings.

While in the field, I learned to make timely decisions that affected the cost and timing of the projects. I learned to work efficiently with multiple groups to include design, plant maintenance, plant operations, construction contractors, equipment suppliers, and management. Most problems encountered in the field are immediate with little or no warning and have to be addressed properly to keep the project moving.

I learned what decisions I could make and inform those affected after the decisions were made and which decisions had to be made by some other group such as the constructor contractor or management. I learned to provide my best assessment of the problem and what could be done to solve it and wait patiently for the decision to be made. In any project, there is a balance that must be reached between the cost of any change and any potential delay to the project. When changes are first developed in the design office, it usually affects a handful of engineers and designers. As the design is incorporated into the project and other disciplines are involved, it may affect a few dozen project team members. When the change is made in the field, it could affect hundreds of people working on the construction of the project and will likely cause a slip in schedule or an unexpected cost to the project.

Design changes are always better if made earlier in the project timeline. This is why critical changes could affect a project when made in the field. There are changes that cannot always be anticipated, such as finding an underground obstruction when attempting to install a foundation in the field. I learned to place an immediate priority on issues that arose in the field and tried to provide solutions as soon as possible to minimize the effect on the project.

All the knowledge and expertise gained during my career I tried to pass along to my peers and especially to the people working for

ENGINEERING

me in all my positions and responsibilities. Before I attained a supervisory role, I shared my experiences with my coworkers, and I also gained knowledge from them on what they were doing. When I first started as an engineer, I often discussed projects with designers and drafters, and even though they did not have the engineering technical background, they did have more knowledge of what would work and what would not work.

I owe a lot of my early knowledge to those around me and would not have been as successful without their help. When I first began to function as a first-line supervisor, I tried to do everything myself. This is typical for anyone as a first-line supervisor, and I was no different. It soon became obvious that I needed to rely on those working for me, and as soon as I began to rely on them, it made the work much easier. I still supervised the work but allowed each of them to take on as much work and responsibility as each could handle. I often said I succeeded by hiring and maintaining good capable people working for me.

If things had gone wrong, it would be my responsibility for the group. If things went well, I was okay with taking a little bit of the praise for hiring and keeping good engineers and designers as long as I praised them for their work and let everyone know that my employees were responsible for the good outcomes. During my career I felt like I always had my employees' back, and they certainly had mine. There were some ups and downs, but overall, it was a very good relationship with those who worked for me.

I have included in the Appendix section some of those things I developed and used during my career to aid the engineering and construction groups in the performance of our tasks. These aids are a small sample of the things used during my career. Many of the guides were specific to a particular company and tasks and were developed as a part of the company documentation and are not included for obvious reasons. Those included are more of a general nature and were not specific to any particular company or restricted documents. It is my hope and desire that anyone following these aids will consider them as general guidelines to be modified as needed to suit their particular needs.

GERALD W. MAYES, PE, RETIRED

It is my aim to present my view of the engineering profession from my perspective by sharing my experiences and knowledge with the reader. I have enjoyed remembering and putting my thoughts into this book and sincerely hope that all who read these pages gain insight and enjoyment from reading it as much as I enjoyed writing it.

APPENDICES

The following appendices are a few examples of different classes and procedures I used throughout my career to instruct and guide those in my group to a better understanding of the work to be accomplished. Some of the guides were developed early in my career and may have to be updated to reflect current codes, regulations, and requirements. Most of these were developed during my later years and reflect knowledge and experience gained that was presented for the benefit of others. Some of the classes, such as the constructability class, were very detailed and will not apply in all cases but was presented as a wide choice of options from which to consider.

The procedures for engineering and checking and the description of a typical petro-chemical plant will need to be modified to fit a particular company's internal requirements and is presented as a typical example. The first-time design class was prepared for my design group when I realized that a number of them were struggling with how to do something for the first time. Enjoy them, especially for those who like to have an outline or guide when taking on a task or new job.

APPENDIX A

Petro-Chemical Plant Design

GERALD W. MAYES, PE, RETIRED

Introduction to Petro-Chemical Plant Design for Civil/Structural/Architectural Engineers and Designers

Typical chemical plant operation (may be batch or continuous)

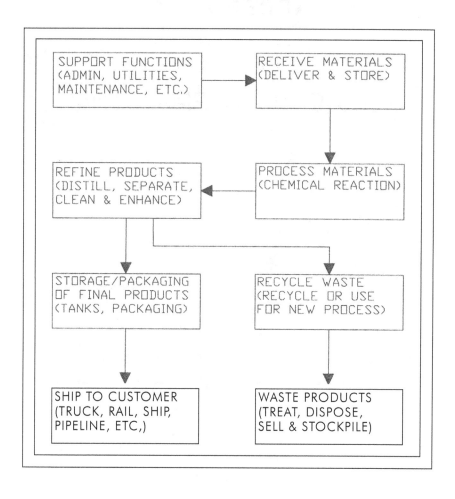

ENGINEERING

Introduction to Petro-Chemical Plant Design for Civil/Structural Engineers and Designers

I. Introduction

Petro-chemical plants have their own unique characteristics and methods of development relative to civil/structural engineers and designers. Unlike the classical design methods taught in school that concentrate on commercial buildings and facilities that house and support numerous personnel, the typical petro-chemical industrial plant may contain many open structures or closed buildings whose function is to support equipment, utilities, piping, and electrical and control wiring and may only contain a few people who are in the structures or buildings periodically to perform a check or function.

Some facilities such as administrative buildings, control rooms, maintenance buildings, and warehouse or storage facilities are similar to conventional buildings from the outside perspective and may contain assigned full-time personnel but are often very different on the inside due to their function. Petro-chemical plants also utilize multiple-level open pipe racks and pipe bridges to connect different structures and storage areas with a function of supporting process pipe, conduit, cable trays, utility pipe, etc. Much of this piping and electrical conduit may also be routed underground between areas.

These plants usually have utility areas consisting of compressors, cooling towers, water tanks, furnaces, boilers, and waste treatment facilities. These plants have load/unload platforms for trucks and railroad cars, load scales, truck docks, metering stations, blowdown lines, flare towers, railroad, and road facilities to support the operation of the plant.

II. Types of Design Projects

A. Greenfield Site

This refers to open land or undeveloped land on which a new plant is to be located. This could be land that is a part of an overall plant property and may or may not have been cleared but generally does not have roads, access, or any utilities that can be used by the new plant. Utilities may pass through the property and there may be adjacent roads or railroads, but none have been developed for this particular site. Many times the land has to be cleared of trees and brush, and there may be a need to provide cut and fill to level the site. Drainage may need to be provided for the area. Some new sites are located near existing facilities and only a part of the area is considered as "greenfield."

B. Additions to Adjacent Existing Plant

If the new plant is located adjacent to an existing plant, there may need to be some clearing, cut and fill, and drainage improvements. But there is generally more concern with tie-ins to the existing facility and trying to match coordinates and elevations between the existing and the new. When adding a new plant near or adjacent to the existing, there is always a desire to use existing facilities if possible to reduce cost and time of construction. Items that should be considered are tie-ins to process, sanitary and stormwater drainage, use of utility steam, cooling water, piping, flare lines, and the use of control rooms, maintenance facilities, change houses, electrical substations, administrative facilities, and parking areas. These decisions are generally based on the capacity of the existing facilities to handle additional loads and the size of the extended plant and will be determined separately on each project.

ENGINEERING

C. Modifications to Existing Plants

This type of project is generally located within the established area of the existing plant although some minor expansion of the facility generally occurs. Most expansion is in the form of another platform level or another bay being added to an existing structure or another tank is added to the tank farm or additional equipment is added to an existing support structure level. Sometimes this involves enlarging existing equipment in its current location. Very often these types of projects will add additional piping to the pipe racks and may include a new level in the racks to support the added piping and cable trays.

Sometimes a new project requires that the existing facilities be removed partially or totally before the new facilities are installed. The drawings may be a combination of new CADD drawings and revised old CADD drawings. Major revisions to existing board drawings are usually placed on new CADD drawings. Minor revisions may be revised by hand on existing board drawings at the discretion of the lead engineer and project manager, but this is the exception and not the rule. The amount of rework or new installation required to support the project is determined separately on each project.

D. Minor Revisions to Existing Supports and Structures

Minor revisions to existing supports and structures may require extensive change, or they may only need to be reanalyzed to determine if they can support the modified loads. Generally, minor revisions do not include the addition of supporting utilities or facilities and should be done as revisions to the existing CADD drawings if possible.

E. Combination Projects

Some projects may include partial greenfield sites, additions adjacent to existing plants, and modifications and minor revisions to existing plants and support structures. A project may be spread over a part of an existing plant site and could even involve multiple locations several miles apart. The scope of these projects are determined on an individual basis.

F. Other Projects

1. *Inspection of facilities.* This may include an inspection and report only and may or may not lead to any engineering or design effort to produce drawings.
2. *Analysis of existing structures and foundations.* This may include the analysis and report only and may not involve any engineering or design effort to produce drawings.
3. *Consultation and review of other projects.* This may include review and comments only.
4. *Drawing pickups.* This may include pickups and revisions to CAD drawings only.

G. Roads and Railroads

Roads will include all main entrance roads, entrances to truck docks and load/unload spots, secondary plant roads, maintenance roads, truck scales, and maintenance access areas. Railroads will include all main railroad lines coming into the plant, turnouts, spur lines, railcar scales, and storage yards.

H. Concrete at Grade or Below

This includes any sumps, basins, trenches, drainage paving, or underground storage tanks not included in the drainage areas in addition to area paving, below grade truck ramps, curbs, pedestals, and thickened slabs, etc. Generally, concrete no more than one foot above grade will be included.

ENGINEERING

I. Concrete Above Grade

This includes any dike walls, concrete structures, concrete columns or beams in a structure, elevated floor slabs, above grade truck ramps, building walls, roofs, pipe racks, retaining walls, concrete tanks, etc.

J. Main Structural Steel

This includes all major buildings, pre-engineered buildings, process structures, equipment sheds, major equipment supports, administrative buildings, change houses, boiler houses, substations (MCC), maintenance buildings, and control rooms that are framed out of structural steel.

K. Structural Steel Pipe Racks

This includes main pipe racks, pipe bridges, miscellaneous pipe supports, and cable tray supports and may include pipe chases in buildings or structures.

L. Miscellaneous Structural Steel

This may include minor equipment platforms, load/unload platforms for trucks and railcars, barricades, handrails, guardrails and posts, bumpers, miscellaneous supports for electrical/instrument equipment, stairs, stiles, crossovers, and other miscellaneous supports.

M. Architectural

1. Siding/roofing on steel process structures and on boiler/compressor houses
2. Control room architectural
3. Substation architectural
4. Maintenance building architectural

5. Change house architectural
6. Load/unload spots architectural
7. Miscellaneous sheds and covers architectural
8. Building signage and safety equipment—exits, no exits, safety equipment, fire extinguishers, controlled access, maximum occupancy, alarm systems, and other requirements for *Life Safety Code*.

N. Modular Design

This includes equipment, structural steel, piping, electrical power/lighting/grounding, instrumentation and controls, temporary bracing, lifting lugs, grating, stairs, handrails, junction boxes, safety showers, etc.

1. All disciplines must be involved.
2. Critical dimensions are out to out and total length as shipped including preattached lifting lugs.
3. The center of gravity, total shipping weight, and the weight per lift point should be provided for lifting shipping and setting of the module.
4. Each separate module must be analyzed for lifting and shipping and should be analyzed as a part of the overall structure.
5. Connections between modules must be analyzed for loads and stresses applied to the connection when considered as a part of the full structure under design loads and conditions.
6. Temporary members may be used in the shipping and lifting analysis but should be removed from the composite structural analysis if the members are scheduled to be removed after installation of the individual modules.

O. Specialty Items

1. Pre-engineered buildings—warehouse, storage, and maintenance buildings

ENGINEERING

2. Pre-engineered building module—substations and control rooms
3. Load cells
4. Nuclear level devices
5. Transformer containment
6. Lifting lugs
7. Monorails and maintenance lift points
8. Bridge cranes and gantry cranes
9. Maintenance access and opening in floors and walls
10. Critical lifts

P. Final Paving and Grading

1. Compaction of soil and use of engineering fabric
2. Areas designated as maintenance or access
3. Final drainage—sheet flow, swales, ditches, underground pipe, catch basins, manholes, and culverts.
4. Landscaping—plants, trees, retaining walls, ground cover
5. All areas to be covered with concrete, asphalt, gravel/crushed limestone, or seeded to provide grass cover to protect against erosion.
6. Parking areas—paint striping, parking lot curbs, handicapped parking, ramps, sign posts, traffic control signs.

Q. Field Support

1. Answer questions on design intent
2. Interpret drawings
3. Resolve problems arising from unexpected circumstances
4. Observe construction to ensure proper installation

R. Subcontracts

1. Geotechnical investigations and reports
2. Boundary and topographical surveys
3. Pre-engineered buildings

4. Turnkey equipment packages
5. Environmental treatment systems
6. Steel detailing and fabrication
7. Reinforcing bar detailing and fabrication

S. Project Close

1. Complete all calculations including those that were required to be confirmed.
2. Clean up purge and file all drawings issued for construction.
3. Clean out and remove all preliminary drawings, sketches, etc. that are not a part of the project.
4. Prepare drawings and calculations for final transfer into project storage.
5. Clean computer files of all excess files not a part of the project.
6. Make final reports, evaluations, etc.

III. Petro-Chemical Plant Components

A. Administrative Areas

This includes administration offices, change houses, control rooms, lunchrooms, etc.

B. Maintenance Areas

This includes maintenance buildings, repair shops, equipment warehouse, etc.

C. Utility Areas

This includes substations (MCC), compressors, water towers/tanks, cooling towers, furnaces, flares, blowdown pits, firewater system, etc.

ENGINEERING

D. Transportation Areas

This includes roads, railroads, truck docks, boat docks.

E. Load/Unload Areas

This includes truck spots, railcar spots, metering stations, truck weigh scales, railcar weigh scales, etc.

F. Raw Material Storage/Supply

This includes storage tanks (tank farm and individual), warehouse storage, railcars, trucks, storage yard, drums, pallets, tote bins, pipeline, etc.

G. Process Areas

This includes reactors, heat exchangers, distillation columns, pumps, drums, evaporators, centrifuges, dryers, separators, blowers, piping systems, instrumentation, valves, hoppers, conveying systems, electrical power, electrical grounding, lighting, control wiring, scales, insulation, fireproofing, coating systems, etc.

H. Product Handling

This includes conveyor systems, pneumatic systems, packaging systems, separation, stripping, piping systems, pipe racks and supports, electrical power, electrical grounding, lighting, instrument controls, etc.

I. Storage Areas

This includes warehouses, tank farms, storage tanks, pipelines, railcars, trucks, etc.

J. Waste Products

This includes material that can be used as a raw product for a different process that can be piped or delivered by truck or rail to an available customer. It also includes material that can be shipped or piped out from the plant for disposal or treatment. It can be released into the air or released into waterways if properly treated to meet the requirements of discharge from the plant. Some waste products are recycled through the plant and reused. Others may be hauled away in dump trucks or bins for disposal in landfills if meeting the requirements of disposal.

IV. Civil/Structural/Architectural Design

A. Site Work

This includes grading of the site, cut and fill, clearing and grubbing, silt fences, construction drainage, construction roads, control of erosion, lay down areas, temporary construction facilities, temporary power, temporary sewer connections or disposal, water supply, etc.

B. Pile Design

This includes steel H piles, steel pipe, precast concrete, and timber piles that are all driven piles. Drilled piles include concrete cast-in-place with or without underreams and auger cast piles. Sometimes combinations of steel pipe or steel mandrel piles are driven and filled with concrete. Micro piles may also be included.

C. Concrete Foundations

This includes mat foundations, strip foundations, ring wall foundations, and spread foundations. Concrete foundations can also be thickened concrete slabs on grade for light loads.

ENGINEERING

D. Structural Steel

This includes steel structures, buildings supporting equipment, pipe supports, and miscellaneous equipment supports.

E. Specialty Buildings

This includes maintenance and storage facilities, control rooms, motor control centers, and administrative buildings.

F. Underground Drainage

1. *Process drainage.* This includes drainage from process areas and equipment drain hubs, lab sinks, load/unload spots, and tank farms (if contaminated) and could include drain lines from cooling towers and wash down drains from packaging areas and may contain main trunk lines, collectors, manholes, catch basins, hubs, and cleanouts.
2. *Sanitary drainage.* This includes sanitary drains from restrooms located in the administration building, the maintenance building, warehouses, and process areas and may contain main trunk lines, collectors, manholes, and cleanouts.
3. *Stormwater drainage.* This includes drainage from building roofs, roadways, railroad tracks, area paving, lay down areas, parking areas, open areas inside the plant boundaries, and process and tank farm areas that have been checked and been determined to be clean before allowing to flow into the stormwater drainage system. Some plant sites will require that all storm drainage be collected in a retention pond for holdup to allow the gradual release of water into the surrounding waterways.

G. Other Underground Installations and Obstructions

This includes firewater systems, cooling water piping, electrical duct bank, water supply pipes, gas pipes, electrical grounding grids and cables, and in some cases, process pipe. All these potential underground obstructions in addition to process, sanitary, and stormwater drainage systems will have to be located and avoided when designing and installing foundations.

APPENDIX B

Engineering Procedures

Engineering Procedures

Proposals

1. Respond to request for proposals.
2. Conduct site visits and gather information for proposal if required.
3. Provide time required for preliminary, material, and cost estimate as required.
4. Determine manpower availability for proposed project.
5. Determine deliverables, methods of design, and software programs to be used for the project.

Award of Front End Loading (FEL)

1. Complete items in "Proposal" section if awarded FEL work without proposal.
2. Provide estimate of manpower to complete FEL phase and detailed design phase for each discipline and group.
3. Provide preliminary cost estimate for the project including equipment, materials, and manpower.
4. Develop project execution plan including scope of work for all disciplines and groups, and provide a specific list of deliverables for each discipline or group.
5. Develop discipline design criteria for the project.
6. Develop tentative project schedule.
7. Develop equipment list for project.
8. Prepare list of long delivery equipment.
9. Develop and issue PFDs and P&IDs for design if not already provided by client.
10. Develop piping line list.
11. Develop equipment arrangement drawings and plot plan and issue for design.
12. Develop electrical one-line drawings and electrical equipment list.
13. Develop instrument list.

ENGINEERING

14. Develop cable tray routing for electrical and instrument disciplines.
15. Prepare major pipe rack layouts, widths, and levels.
16. Prepare preliminary drawings for major buildings and structures.
17. Prepare preliminary layout of tank farms and storage areas.
18. Request geotechnical report if needed.
19. Request site survey if needed.
20. Set up kickoff meeting, and schedule regular project/design meetings.
21. Prepare set of preliminary drawings and calculations for delivery to the client.
22. Prepare a preliminary material takeoff by discipline for delivery to the client.
23. Receive firm quotes for major pieces of equipment.
24. Determine required deliverables to be provided by third party as a turnkey installation and develop package for turnkey bid.
25. Provide complete deliverables package to client upon completion of FEL phase.

Award of Detail Design

1. Hold department-wide kickoff meeting for detailed design.
2. Present approved project execution plan to project team for specific requirements, guidelines, and deliverables for each discipline or group.
3. Begin foundation design and preliminary structural steel design for civil group.
4. Begin piping design and isometric drawings based on P&ID-approved drawings.
5. Begin electrical design of transformers, switchgear, grounding, lighting, and distribution systems.
6. Begin selection and design of instrumentation and controls.

7. Begin mechanical group preparation of equipment specifications and requirements for major equipment to be designed and fabricated by outside vendors.
8. Begin selection of valves, instruments, controls, and specialty items per P&IDs.
9. Begin mass material balance design of process systems in accordance with P&IDs.
10. Submit equipment and material specifications to client for approval.
11. Track budgets, specifications, line lists, electrical lists, and drawing progress.
12. Update progress at weekly project meetings to include work completed since last meeting, work currently in progress, and two- or three-week look ahead work planned. Also include information needed with group or person to provide information and a requested date. Follow up on information dates past due with e-mail correspondence request.
13. Utilize field trips for measurements, data collection, or review with site personnel as needed.
14. Schedule internal model reviews as needed with design groups and project management to identify areas needing improvement, and provide action item list and date requested for each group to accomplish the action items.
15. Schedule external model reviews with the client, project management, and discipline representatives after internal reviews at a designated time in the project such as 30 percent, 60 percent, and 90 percent of design completion as indicated in the approved project execution plan.
16. Complete all discipline drawings, calculations, specifications, lists, requisitions, and reviews as indicated in the project execution plan and self-check each discipline item.
17. Utilize discipline checklist to complete drawings, calculations, etc.
18. Provide all discipline deliverables for formal check per discipline and company requirements and include discipline checklist.

ENGINEERING

19. Provide all discipline deliverables for interdiscipline check.
20. Pick up or address all comments received internally from group or other disciplines.
21. Issue deliverables to client for approval, IFA.
22. Pick up or address all comments from client.
23. Provide PE stamp on all required drawings, documents, and deliverables, and issue complete deliverable package to client for construction, IFC.
24. Clean up files, transmit original drawings and required documents to client.
25. Provide services as required and listed in project execution plan such as responding to RFIs (request for information), revisions due to unforeseen field requirements, revisions due to vendor drawing changes and additions, and review of structural steel fabrication drawings if requested and listed as a part of the project deliverables.
26. Be prepared to perform additional work at the client's request per approved change orders.

Additional Administrative Functions

1. Fill out electronic time sheets on a daily basis and submit on a weekly basis at the end of the work week with the correct approved charge codes.
2. Submit timely expense reports in accordance with company procedures.
3. Keep all projects documents secure when not at your work area.
4. Utilize the shred bins for project and sensitive documents when no longer needed for the project.
5. Check e-mails on a regular basis for updates and requests on projects.
6. When not in the office and during periods of high project activity, check your e-mail from your smartphone or home computer to keep informed and stay on top of activities and requests.

7. When out of the office, make sure your immediate supervisor or other appropriate personnel know you are out and provide a means of contact such as a cell phone.

APPENDIX C

CSA Department Checking Procedures

GERALD W. MAYES, PE, RETIRED

CSA Department Checking Procedures

I. Definitions

Checking. Checking is the process of verifying that the correct setup, engineering and design procedures, and representation of the engineered and designed product have been followed and that drawings (before being issued for fabrication and/or construction) contain the proper concept, views, dimensions, notes, and references to convey fully to the client what is required and expected to be done.

Procedures. Procedures are instructions on how to perform a task that has been compiled, approved, and published. Procedures that are listed in the engineering plan are to be followed during the duration of a project.

Design criteria. The design criteria is a list of values that are to be used as a basis of the design of a project. These items may be specified by the client, required by code or regulation, or developed from experience.

Guidelines. Guidelines are suggestions as to how a project is designed or developed.

Specifications. Specifications are requirements on how a project is designed or developed and addresses the specifics of what is to be designed, fabricated, installed, inspected, or delivered. Specifications, if listed as a part of the design package, must be followed on the project.

Standards. Standards are typical methods and drawings of designs that have been utilized in the past and have been previously checked and verified. Standards may be used without checking if they are suitable for the application for which they are being used.

ENGINEERING

Building codes. Building codes are model codes that provide minimum requirements to safeguard the public health, safety, and general welfare of the occupants of new and existing buildings and structures.

Life Safety Code. The *Life Safety Code* specifically addresses the design of buildings and structures as related to life safety from fire, emergency, and other related hazards encountered in buildings and structures. The *Life Safety Code's* primary concern is life safety and not necessarily the protection of property.

Technical codes. Technical codes such as AISC and ACI, NDS for wood, etc. covers the proper design and construction of various buildings, systems, or components. Technical codes are often included by reference as a part of a building code.

Technical standards. Technical standards such as NFPA, ASCE/SEI-07, AREA, etc. provide technical details, guidelines, and specifications related to the particular industry or material they represent.

Regulations. Regulations, such as OSHA and ADA, are generally referred to as the *Code of Federal Regulations* and are required and enforced by federal law.

Calculations. Calculations are the required analysis and justification for the design of components and systems and include hand calculations, design spreadsheets, and computer software programs. Calculations are based upon design criteria, codes, regulations, standards, and knowledge and experience of the design engineer.

Methods. Methods generally refers to the way in which calculations are performed or the way in which the design is developed, such as hand calculations, computer analysis programs, client requests, and similar past design and experience.

Results. Results refers to the output of the computer analysis programs, spreadsheet results, hand calculation results, or the outcome of a design method or decision.

Conclusions. Conclusions are the outcome and evaluation of the results and a determination of what should be utilized in the design of the components or systems being analyzed.

Drawings. Drawings are formal representations of the components or systems to be fabricated or installed and are placed on standard-size layouts with title blocks and signature blocks. Drawings may be on paper, as an original with signature and PE stamped, or it may be as a PDF copy or an electronic copy.

Sketches. Sketches are informal drawings that may not necessarily have a standard-size layout, a title block, signatures, or PE stamps. Sketches are often used as a draft of a potential drawing to convey an idea or a possible scope of work that may be converted into a drawing at a later date or may be discarded when no longer useful.

CADD platform. The CADD platform refers to the CADD system utilized to produce models, drawings, takeoffs, etc. for the assigned project and includes AutoCAD 2D and 3D with associated ProSteel software, MicroStation 2D and PDS FrameWorks, SolidWorks, etc.

Models (CADD and analysis). CADD models are a 3D representation of a structural system that contains intelligent information but generally does not contain dimensions, text, or notes. CADD models are used as cross-references to other disciplines, interference checks, location of center of gravity, weight of the structure by component type, etc. Analysis models (STAAD.Pro) are representations of the structural components that make up the structural system being analyzed.

ENGINEERING

Model space/paper space. Model space incorporates the entire structure or system in 3D and is generally drawn in true dimensions and coordinates. Paper space is a view of that model or a portion of that model in 2D that is represented inside of a viewport or multiple viewports located on a 2D drawing.

Analysis program. Analysis programs include STAAD.Pro, Mathcad, spMats, Mat3D, Foundation3D, Excel spreadsheets, etc. that utilize computer analysis methods to solve complex structural problems and provide output for the design of components and systems. The analysis programs must be set up with the required geometry, specifications, and parameters for the proper solution of the problem.

Input. Input is the required loads, load combinations, and any adjustments to the program geometry, specifications, or parameters that are required for a specific analysis.

Output. Output is a listing of the reactions, deflections, code check, unity check, etc. for a given set of input values.

Model walk-through/review. A model/walk-through review is a multidiscipline review of a system model using a review software, such as Navisworks, to move through the combined discipline model looking for interferences, omissions, and improvements in order to gain a better understanding of how the completed installation will look and function. Comments from the model review are noted, and each discipline has a list of modifications to complete before the design packages are issued for fabrication and/or construction.

Engineering checking. Engineering checking includes checking of the calculations and the adequacy of the component sizes, dimensions, arrangements, connections and provided views/details as represented on the 3D model and/or the 2D drawings before issuing for fabrication and/or construction.

Designer checking. Designer checking includes the model setup, number, and types of view selected for drawings along with proper use of dimensions, notes, text, scale, layout, and completeness of the 3D models and 2D drawings prepared to issue for fabrication and/or construction.

Constructability. Constructability is simply determining if the information conveyed on the drawings can be fabricated and installed safely and efficiently in the field. This check includes welded and bolted joints, installation of underground items that may cause problems with adjacent structures, location of cranes to set equipment, steel, concrete pours, piping, etc.

Interferences. Interferences include anything that prevents the proper operation of the facility or system such as piping routed through steel members, valves with actuators that hit steel or stick out in walkways, or conduit that is run in areas that are required to be kept clear for access or equipment removal.

Interdiscipline check/review. The interdiscipline check is a check of a discipline set of drawings that are routed through all disciplines that are affected by the design and drawings for a check on concept, constructability, interferences, and comments. The set of drawings has a stamp for routing through the disciplines with a sign-off when the check is complete. The drawing set continues through each discipline until all have been checked and signed off, and then it is returned to the original discipline for pickups before issuing the drawings for fabrication and construction.

Vendor drawing check. The vendor drawing check is a review of vendor drawings submitted for approval of equipment, packaged systems, or products provided by others to be used in the completed design package. The CSA group often checks weights, dimensions, layout, clearances, support lugs, platforms attached to equipment, attachment plates welded to equipment, seismic and wind loads designs, etc.

ENGINEERING

Concrete rebar fabrication check. Concrete reinforcing is sometimes checked if the fabricator provides a rebar fabrication schedule. CSA would look at proper rebar size, length, hooks, and number of bars to see if it covers the drawings being produced for the project.

Structural steel fabrication check. Structural steel fabrication drawings are checked to ensure that the structural steel members delivered to the site have been detailed and fabricated per the drawings, structural steel notes, and specifications for the project. The level of detail varies with projects and may be a general overall check to see if all the steel members have been accounted for. Or it may involve a very detailed checking procedure that includes the design of the connections and accounting for each and every note, dimension, weld, bolt, nut, and washer to be delivered for the project.

II. Calculation Checking Requirements for Engineers

A. Check calculations for adherence to procedures (those referenced in the engineering plan, on drawings, and in calculations).

B. Verify/check design criteria—CSA and client design criteria.

C. Verify/check code and regulation requirements:

1. OSHA
2. *Life Safety Code*
3. IBC
4. ACI
5. AISC
6. AREA
7. ADA, etc.

D. Verify analysis method used:

1. STAAD.Pro
2. spMats
3. Mat3D
4. Foundation3D
5. Excel spreadsheet
6. Mathcad
7. Hand calculation, etc.

E. Verify geometry of the system being analyzed:

1. TOS elevations
2. Bay spacing and number of bays
3. Dimensions to/coordinates of equipment based on plot plan or equipment arrangement
4. Top of concrete/top of grout
5. Spacing of pipe supports
6. Pipe rack levels
7. Pipe rack widths
8. Location and size of miscellaneous platforms
9. Type, location, and size of foundations/piles

F. Check analysis input for loads and load combinations:

1. Dead loads (structure, equipment, etc.)
2. Live loads (floor, roof, equipment contents, etc.)
3. Wind loads (N, S, E, W and diagonal)
4. Snow and ice loads
5. Seismic loads
6. Thermal, anchor, maintenance, construction, etc. loads
7. Secondary analysis loads
8. Combination loads (IBC, ASCE, PIP, client, CSA, etc.)
9. Check to make sure loads are properly applied to the structure/foundation and are of the correct values

ENGINEERING

G. Verify analysis files are set up properly—correct form and order.

H. Verify units, member sizes, member releases, and support conditions match STAAD.Pro input.

I. Verify correct parameters selected for STAAD.Pro analysis program:

   1. Strength and modulus of structural steel
   2. Braced lengths
   3. Member type
   4. K values
   5. Beta angles
   6. Code to be checked by
   7. Specify required output
   8. Reactions at support joints
   9. Member stresses
   10. Deflections
   11. Code check
   12. Weight of structure by member size

J. Verify correct parameters selected by foundation analysis program

   1. Length, width, thickness, and element size of foundation
   2. Strength of concrete, strength of reinforcing steel, cover on reinforcing steel, and soil strengths or spring constants

K. Concur with conclusions and results from analysis

L. Refer questions and concerns to responsible engineer for consideration and rework if needed.

M. Provide signature and date as reviewer on calculation cover page.

N. Provide PE stamp on cover page if required.

III. 3D Model Checking Requirements for Designers

A. Check proper setup of CADD model using standard templates/guidelines.

B. Check cross-reference with other models (piping, mechanical, etc.).

C. Check equipment size and location versus steel and concrete models.

D. Check for interferences between discipline models.

E. Check for interferences with existing equipment, structures, piping, etc.

F. Check for the addition of miscellaneous pipe supports to piping model.

G. Check for change in equipment size or location in the mechanical/piping models.

IV. Drawing Checking Requirements for Engineers and Designers

A. Use correct drawings procedures and standards for setup

B. Drawings should be separated by drawing types

C. Check drawings for appropriate scale

ENGINEERING

D. Check use of subtitles for views

    1. Plan
    2. Elevation
    3. Section
    4. Detail
    5. Profile
    6. Enlarged plan
    7. Self-described

E. Check dimensions

    1. Dimensions match calculations
    2. Strings of dimensions equal total
    3. Appropriate dimension style
    4. Check against model

F. Check orientation from view to view for consistency

G. Notes and callouts on appropriate views

H. Check for strikeovers and crowded views

I. Check for text or drawing outside of viewport

J. Reference drawings on views

K. Reference drawings on drawing

L. Verify title block information

M. Verify coordinates, grids, and north arrow

N. Check steel connection details

    1. For constructability

2. For proper type of connection
3. Proper notes and callouts
4. Insure that welds can be made

O. Check concrete details

1. Will anchor bolt fit inside pedestal reinforcing?
2. Are there adequate numbers of horizontal ties for the vertical reinforcing?
3. Can the load be transferred from the anchor bolts to the vertical reinforcing?
4. Is there sufficient development length for the reinforcing?
5. Are anchor bolt projections and threads adequate?
6. Is the grout of sufficient depth?
7. Is the vertical steel anchored sufficiently in the foundation?
8. Will a mud slab be required?
9. Will shoring be required for the excavation?
10. Does the anchor bolt size, location, and projection match the structural steel baseplate?

P. Check structural steel framing

1. Is it stable?
2. Do working platforms have proper access via stairs or ladders?
3. Do the platform handrails miss the bracing?
4. Are safety gates used at all ladders?
5. Are stairs compliant with OSHA/*Life Safety Code*?
6. Are work points located and called out on the drawings?
7. Are all member sizes shown in the proper view?
    a) Horizontal beams, girders, and bracing are shown on the plan view
    b) Columns and vertical bracing are shown in the elevation view

ENGINEERING

      c) Baseplates, end plates, gussets, etc. are shown and called out in detail or schedule views and may be shown in elevations

      d) Some items may be called out in notes or provided in tables

Q. Civil/Structural/Architectural Checklists

1. General
2. Pilings and drilled shafts
3. Site work
4. Area grading/paving/drainage
5. Underground piping
6. Concrete
7. Structural steel
8. Architectural
9. Revisions

R. Steel Fabrication Checking

1. General
2. Correct member sizes and quantity
3. Orientation of columns, girders, beams, and braces
4. Conflict of handrails and vertical bracing
5. Correct dimensions
6. Connection design
7. Gusset plates and clip angles
8. Base/cap plates
9. Bolt hole size, number, and location
10. Constructability of structural steel
11. Pipe support and pipe hangers
12. Anchor points

S. Concrete Reinforcing Checking

1. Reinforcing specification and strength

2. Size reinforcing
3. Reinforcing spacing/number of reinforcing bars
4. Lap splice lengths
5. Reinforcing embedment lengths
6. Reinforcing hooks and anchors
7. Size and number of ties at pedestals
8. Reinforcing chairs
9. Other embeds

T. Vendor Equipment Checking

1. Weight, empty and operating
2. Location and magnitude of dynamic loads, vibrations, etc.
3. Agitator loads
4. Extended footprint for equipment access
5. Overall dimensions
6. Number, size, and location of protruding nozzles, manholes, etc.
7. Manway davits and jib hoists attached to equipment
8. Access location points
9. Ladder, platform, and stairs location and attachment to equipment
10. Insulation thickness
11. Support location and number of supports
12. Size of support lug, baseplate, support frame, etc.
13. Number, size, and location of anchor bolt holes
14. Number, size, type, and location of guide supports
15. Equipment design based on specified wind and seismic loads per code

U. Intra-Squad Checking

1. Equipment numbers
2. Coordinate or dimensional location
3. Location of pipe supports and piping anchors

ENGINEERING

4. Elevation of top of grout versus bottom of equipment baseplate
5. Elevation of top of grout versus face of flange, pipe centerline, etc.
6. Elevation of top of steel support or platform versus bottom of equipment baseplate or bottom of pipe/pipe shoe
7. Verify access ways are unobstructed with piping, conduit, valves, braces, supports, structural steel framing, etc.
8. Elevation, size, levels in pipe rack match elevations and locations of pipe runs and piping turnouts from rack
9. Verify location of ladders, stairs, ramps, etc. are not obstructed
10. Constructability of the facility

APPENDIX D

First-Time Design

ENGINEERING

The New Design Concept

(Or how to do something for the first time and get it right even though you didn't know what you were doing before you started.)

A. Assignment of a Task

   1. Why are you giving it to me?
   2. I've never done this before.
   3. I'll probably miss something important.
   4. Will someone check my work?
   5. I may take too long to do the work.
   6. I'm not trained to do this.
   7. What if I fail?
   8. Can someone help me?
   9. Can I watch someone else do the work and learn that way?
   10. I'll do the best I can.

B. Information Needed to Do the Work

   1. Do I know the scope of work?
   2. Do I have the information required to do the work?
   3. What assumptions do I have to make?
   4. Am I qualified to make these assumptions?
   5. Can I get someone else to verify my assumptions?
   6. Can I count on others to provide information to me that they have promised in a timely manner?
   7. What if I make a bad assumption?
   8. What if I don't understand the information when I receive it?
   9. Can I ask for additional information?
   10. I'll use the information I have with the assumptions clearly stated and do the best I can to complete the work.

C. How Do I Start the Work (Start at the Beginning)

1. Look for the obvious first. Determine what deliverables are needed and in what order they need to be completed.
2. Then divide the work into groups or tasks to accomplish the goal, and start on the first task.
3. Spend more time on those tasks that affect the largest portion of the overall work if time is limited.
4. Spend less time on those tasks that affect the smallest portion of the overall work if time is limited.
5. Document the work so that others can follow your thought process.
6. Ask for help if you need guidance to get started in the right direction.
7. Don't ask for someone else to do your work for you because you will not learn as much as when you do it yourself.
8. Each task you do should build toward the overall goal.
9. Try on your own before asking someone else to show you how.
10. Remember, this is your task. So get started, and do your best.

D. How Do I Know if I am Accomplishing My Goals?

1. As you complete each task, review your work and see if you are still working toward the final goal.
2. Ask the opinions of others from time to time, and consider their advice.
3. Look at the percentage of work you have accomplished and how much of your budgeted time you have spent. Are you still on track, and can you finish within the time allotted?
4. Do the answers to the first tasks provide input for the tasks to follow?
5. Stop and consider if the answers you are getting are logical and within expected limits.

6. If you are not sure, ask someone else if the answers seem reasonable.
7. If your assumptions are way out of line, you may need to go back and correct your assumptions and rework your tasks up to the present location and then reevaluate your answers to see if they are reasonable.
8. If you are not sure about your assumptions, this may be a good spot to evaluate alternates based on different assumptions and complete your tasks and then evaluate the various options.
9. Apply the duck test. (If it looks like a duck, walks like a duck, and quacks like a duck, it must be a duck.)
10. Continue doing a good job until you complete the work.

E. What Tools Shall I Use to Accomplish the Work

1. Hand calculations may be appropriate for some work but must be checked very close to ensure that it is correctly done.
2. Hand sketches may also be appropriate for some work but may not be easily modified and usually are not to scale and may not be easily reproducible to include in client documents.
3. Calculations produced by using software such as Mathcad or Excel spreadsheets can be accurate and very presentable to the client but should always be checked to make sure the correct input is used and the correct software with embedded formulas are appropriate.
4. Analysis of structural systems using software such as STAAD.Pro or pcaMats must be reviewed to make sure that all parameters, releases, and input are correct.
5. Make sure you are using the proper codes, regulations, and design criteria for the project.
6. Always understand what the software does and be able to reproduce it in a hand calculation if necessary.

7. Never depend upon the output of a computer program when you don't know how the answer was obtained.
8. Understand what the various parameters and boundary conditions are, and make sure they are appropriate for what you are using.
9. You are not expected to produce designs or calculations involving multiple iterations that might take weeks to accomplish by hand, but you are expected to know what the software does and why it is doing it in order to use it to support the work you are trying to accomplish.
10. Software is your friend. Use it, but don't abuse it.

F. What Do I Do When I am Complete

1. Review your work for errors, and make sure you understand what you did.
2. Have a peer or supervisor review you work package and provide comments.
3. Determine if you have provided all the deliverables required for the work assigned.
4. Are the deliverables in the proper form?
5. Have the deliverables been saved in the proper form for filing?
6. Have my manual files and computer files been cleaned and purged if needed?
7. Have my files been properly located if they are a part of an ongoing project?
8. Do I need my supervisor to sign off or approve my work?
9. Do I need to return any of the information I used to accomplish the work?
10. Report to my supervisor that I am complete with the assignment and request additional work be assigned to me.

G. Was It a Good or Poor Job

1. Did I accomplish my goal?

2. Did I provide the requested deliverables in the form expected?
3. Did I do it within the time allotted?
4. Did I do my own work?
5. Did I seek guidance from others?
6. Did I properly use software tools?
7. Did I make it clear so others can follow my work process?
8. Did I have it checked and approved by my supervisor?
9. Did I gain experience and knowledge in this type of work?
10. If the answer is yes, then pat yourself on the back. You did a good job. Now go and learn how to do something else, or continue to improve your skills in this area.

H. What Positive Things Did You Learn

1. You learned that you were capable of doing things you did not know how to do with the proper guidance and instructions.
2. You learned to trust your instincts and use positive reinforcement from others.
3. You learned that sometimes you make mistakes and may err from time to time, but with proper guidance and motivation, you can correct those mistakes and complete the work.
4. You learned that not all output from a computer is valid and needs to be checked and verified. GIGO (garbage in, garbage out).
5. You learned to ask for guidance from others but not to depend on them to do your work.
6. You learned the importance of having the required information before starting a job.
7. You learned to clarify your assumptions before starting the work.
8. You learned to identify your deliverables by name and format before starting the work.

9. You learned that not everyone will provide you with the information required when you requested it and that you must handle this situation.
10. You learned to do something that you did not know how to do before, and you will continue learning more about a particular practice every time you do it from now on.

I. What Improvements Can You Make

1. When you make a mistake, admit it, ask forgiveness, and learn from your mistakes.
2. Defend your actions only to the extent to explain the situation, not to place the blame on someone else.
3. Work toward changing the situation that brought about the mistake so others will not be subjected to the same fate.
4. Work toward improving skills that will prevent mistakes from occurring.
5. Document your work so that a misunderstanding does not lead to a mistake.
6. Understand that you are the key to preventing future mistakes from happening again.
7. Use your experience to guide and caution others.
8. Don't wait until a possible mistake or accident is likely, but be proactive and help to remove situations and causes before they become serious.
9. Try to do a better job each time you do a project than the previous time.
10. Make improvements not just a goal but a continuous attitude.

J. Benefits of Doing a New Job Right the First Time

1. Personal satisfaction
2. Builds confidence
3. Develops skills for the next challenge
4. Prepares you for a promotion or pay raise

ENGINEERING

5. Enhances your capabilities for doing other types of work
6. Impresses your boss (see no. 4)
7. Enhances stature among peers
8. Helps the company you work for
9. Helps you sleep good at night
10. (You fill in the blank:) _____

APPENDIX E

Rule of Thumb Class

ENGINEERING

Engineering Rules of Thumb

Rule of thumb—a broadly accurate guide or principle based on experience or practice rather than theory.

Why Rules of Thumb?

1. To make on-the-spot intelligent decisions
2. To provide reasonable preliminary input for computer models
3. To validate the results of computer models

Sources of Rules of Thumb

1. Collected from peers
2. Discerned through observation
3. Derived

Cautions

1. Rules of thumb are approximations and not always conservative.
2. Results must be verified by more comprehensive investigations.
3. Rules of thumb should never be used as the final determination of a design or member selection for approval or construction drawings.
4. Rules of thumb are to be used for preliminary designs, approximations, and as a starting point for further calculations.

5. Rules of thumb are also very useful for providing estimates before any calculations are made and should be reflected in the estimate with a "+/-" confidence.

Origins

Meaning of the *rule of thumb*: a means of estimation made according to a rough and ready practical rule, not based on science or exact measurement. The phrase itself has been in circulation since the 1600s. The origin of the phrase remains unknown. It is likely that it refers to one of the numerous ways that thumbs have been used to estimate things judging the alignment or distance of an object by holding the thumb in one's eyeline, the temperature of brews of beer, measurement of an inch from the joint to the nail to the tip, or across the thumb, etc.

The phrase joins "the whole nine yards" as one that probably derives from some form of measurement but which is unlikely ever to be definitively pinned down. The Germans have a similar phrase to indicate a rough approximation: "pi mal daumen," which translates as "pi (3. 14…) times thumb."

For purposes of this class, we will select some of the more appropriate ones and divide the rules of thumb into three groups: (1) those derived from accepted equations and calculations and presented in a much simpler and approximate form, (2) those observed from practice and experience, and (3) those based on opinions with a psychological view.

Derived Rules of Thumb

1. For a steel beam of cross-sectional area A in square inches and a weight of Wt in pounds per foot: $A = Wt/3.4$
2. For a steel beam of height d: $I_x \sim d^2 * Wt/20$
3. Using this to determine the section modulus: $S_x \sim d*Wt/10$

ENGINEERING

4. Through graphical observation the relationship between radius of gyration and the height d and the width b of the cross section:
   rx~45%*d
   ry~25%*b
5. Substituting the previous assumptions for section modulus into the formula $M = F_b * S_x$, we get:
   Wt ~5*M/d for an allowable 24 ksi bending stress
   Wt ~3.5*M/d for an allowable 33.5 ksi bending stress
6. Beam depth: beam depth = ½ inch per foot of span
7. Weld shear capacity: For a 1/16-inch weld, one-inch long the shear capacity is approximately 925 pounds. Multiply the weld by the number of sixteenths and by the length in inches to get the total capacity. Example: a 3/16 weld four inches long has a capacity of (925)(3)(4) = 11,100 pounds.
8. A general shortcut for beam design: Instead of calculating beam deflection after selecting the beam, invert the deflection equation and calculate the minimum I value for L/360 (or L/240 or absolute value) deflection limit. Then you can look up both properties and quickly select a beam from the handbook such that the M and I tabulated are greater than your computed values.
9. Another time saver for deflection checks is to take the sum of the moments computed from a combination of loads and convert to an equivalent uniform load and determine the deflection for the uniform load, which should be a reasonable approximation of deflection (or required I).
10. In reinforced concrete design to determine the concrete beam area of steel required for flexure:
    As = Mu/(a*d)
    Where *As* = area of steel required in inches
    Mu = design moment in ft-kips (assume fy = 60 ksi)
    a = variable based on the concrete compressive strength
        (example: use 4.0 for 3,000 psi concrete)
    d = depth to the centroid of the reinforcing steel

11. To design a foundation for a vibrating machine: weight of the footing is equal to three to five times the weight of the machine.
12. When you double the pipe diameter, it gives you five times the flow.

Observed Rules of Thumb

1. In soil mechanics, there is a rule of thumb that actually involves your thumb: To evaluate the compaction of a clay material, press your thumb into the compacted surface. If you thumb sinks all the way in, ~75 percent standard. If you reach the first knuckle, ~90 percent standard If you can dent it only, ~95 percent standard.
2. For right-handed threads, if you turn the bolt or nut in the direction of your folded fingers of your right hand, the linear movement of the bolt or nut is in the direction of your thumb. This is also known as "righty tighty, lefty loosey."
3. How about thumbs-up for approval and thumbs-down for disapproval?
4. Measure twice and cut once. This also applies to many things in addition to cutting.
5. If it looks funny, check it again.
6. If it draws hard, it builds hard.
7. If it ain't broke, don't fix it.
8. Short form specifications:
    a) Use good stuff, do good work
    b) Hammer to fit
    c) Caulk to patch
    d) Paint to match
9. Plumbing: hot on the left, cold on the right, s__t won't flow uphill.
10. Time management for new engineers and designers: 20 percent of your time will be spent on the design concept

and general arrangement, 80 percent will be spent on the details. The 20 percent is often the fun part.
11. There is more than one way to skin a cat.

Opinions

1. It is better to find an approximate solution to an exact problem than to find an exact solution to an approximate problem.
2. When you think you know everything, there is someone who knows a few more things than you. (The key is to find out who that is and learn as much as you can from that person.)
3. The client will always try to alter your design during construction.
4. The brighter the color of your architect's socks, the higher chance you have of being asked to frame something totally beyond the laws of statics.
5. Everything you need to know about civil engineering:
   Sum of $F = 0$
   Sum of $M = 0$
   The rest of your education/experience is learning when and how to use the above.
6. The more I learn, the more I realize just how little I know.
7. If you don't have time to do it right, when are you going to find time to do it over?
8. If something can go wrong, it will go wrong,
9. Rule of thumb for everyone who works:
   a) There's always someone out there who is smarter.
   b) There is always someone out there cheaper.
   c) All the employees think they are underpaid.
   d) All the employers think they pay the employees too much.

10. Use rules of thumb wisely as an aid and preliminary estimating tool, but do not rely on them unless you know where they come from and their potential liabilities. Never use them as a sole source for a design, especially if there is a safety issue involved.

APPENDIX F

Constructability Class

Constructibility

Constructability (or buildability) is a project management technique to review construction processes from start to finish during the preconstruction phase. It is to identify obstacles before a project is actually built to reduce or prevent errors, delays, and cost overruns.

The term *constructability* defines the ease and efficiency with which structures can be built. The more constructable a structure is, the more economical it will be. Constructability is, in part, a reflection of the quality of the design documents—that is, if the design documents are difficult to understand and interpret, the project will be difficult to build.

Design Guidelines

The object of this procedure is to make optimum use of construction knowledge and experience in planning, design, procurement, and field operations to achieve overall project objectives.

A common view of design guidelines involves only:

- determining more efficient methods of construction after mobilization of field forces,
- allowing construction personnel to review engineering documents periodically during the design phase,
- assigning construction personnel to the engineering office during design, and
- a modularization of preassembly program.

In fact, each of these represents merely a part of the optimization process. Yet only through effective and timely integration of construction input into planning, design, and field operations will the potential benefits of optimization be achieved.

The planning and execution phases for a typical major industrial project involve conceptual engineering, detailed engineering, procurement, construction, and start-up.

Construction optimization analysis should begin during the conceptual stage, at the same time as operability, reliability, and maintainability considerations surface. It can then continue through the remaining phases. Planners must recognize that the payoff for optimization analysis is greatest in the earliest phases of a project, growing progressively less (but never ceasing) until the end of the project.

In modern engineering jargon, this process of design optimization is called constructability.

Constructability

Constructability analysis is a form of both value engineering (VE) and value analysis (VA) that focuses mainly on the construction phase.

Constructability decisions are oriented toward:

- reducing total construction time by creating conditions that maximize the potential for concurrent (rather than sequential) construction, and minimize rework and wasted time;
- reducing workhour requirements by creating conditions that promote better productivity or creating designs that demand less labor;
- reducing material costs through more efficient design, use of less costly materials, and creation of conditions that minimize waste;
- creating the safest workplace possible since safety and work efficiency go hand in hand; and
- promoting total quality management (TQM).

Essential Elements of Constructability

Three elements must be present if a constructability program is to realize its full potential.

First, constructability must be viewed as a program that requires proactive attention. The mistaken idea that constructability is a review of completed designs by someone familiar with construction is totally wrong. By the time designs are ready for review, it may be too late to change anything, and if such changes are made, they will be costly. In other words, constructability is a component of planning that must be included in all phases.

Second, constructability is a team effort. Only if the interests of all parties are jointly represented in all decisions will the optimum solution be realized. Reducing construction costs is certainly an important objective, but doing so must not compromise other needs.

Third, constructability must have management commitment and support. The time and resources needed for such a program must be made available if the program is to be a success.

A Constructability Program

While no single program will fit every project, the consensus is that most successful constructability programs have the following elements:

- clear communication of senior management commitment to the program,
- single-point executive sponsorship of the program,
- an established policy and program as well as tailored implementing programs for each project,
- a database compiling lessons learned and examples,
- orientation and training as needed, and
- active appraisal and feedback.

ENGINEERING

Constructability Culture

Constructability works best when it is an accepted part of the way an organization operates. If the subject is given enough emphasis and attention over time, it becomes ingrained within the organization, reaching what can be called a constructability culture. Every staff person must feel part of the system since their input is frequently sought in constructability brainstorming sessions and their ideas are welcome additions to the database.

GERALD W. MAYES, PE, RETIRED

Constructability Concepts

The Management Approach

Recognize that start-up and construction drive engineering and procurement scheduling. Develop a network schedule as early as possible and include engineering and procurement packages in the control schedule. Use contracting and management approaches that promote construction efficiency by utilizing the following:

- On engineering-procurement-construction projects executed on a fast-track basis (overlapping phases), use single management of the total effort from the outset of conceptual engineering.
- Use construction contract packages of a quality that will allow fixed-price bidding as a means of reducing or eliminating the problems associated with changes.
- Do not start on a work package until the availability of all required resources are assured (personnel, materials, and support equipment).
- Work with the owner to use any existing facilities or services rather than creating duplicate ones for the construction period.
- When packaging designs for specialty subcontracting, consider normal jurisdictional lines so that packages logically fit the specialty contractors involved and do not require sub-tier contracting.
- Plan the release of contracts to take advantage of favorable construction weather.
- Provide adequate planning time for contractors and subcontractors in the bidding award process.
- Keep the control schedule at a summary level.
- Ensure the project milestones are reasonably attainable considering both construction and procurement time.
- Do not impose unnecessary hold points for quality checks.

ENGINEERING

- Keep requirements for the owner involvement in the project (such as reviews and approvals) to a minimum.
- Issue instrumentation, piping, and insulation packages as early as possible since these require the most field time to execute.
- Use a contract form that incorporates incentives designed to reduce construction costs. For example, include value engineering/value analysis clauses that provide for sharing in savings provided by adopting cost improvement suggestions made by the contractor.

Ensure project requirements and conditions are understood by dong the following:

- Make certain that field conditions are accurately reflected on design documents.
- Identify all access routes and any limitations on their use.
- Be sure that all parties understand their roles and responsibilities with regard to providing equipment, the use of project areas and facilities, security and gate control, administrative policies, etc.
- Identify disposal areas for excavations, vegetation, and nonhazardous waste.
- If working in or adjacent to an operating facility, identify all constraints that the situation presents.

Design Phase

Emphasize Standardization and Repetition

Standardization and repetition maximize application of the learning curve to the work force, permit volume buying of materials, and simplify purchasing and warehousing as follows:

- Standardize structural members, foundations, bolt sizes, and other components as much as possible.
- Dimension concrete components to take advantage of readily available commercial form sizes.
- Repeat designs throughout the facility (this will reduce design costs while promoting the learning curve effect during construction).

Take maximum advantage of readily available, off-the-shelf materials and components such as the following:

- Maintain access to commercial catalogs of equipment and materials.
- Make maximum use of vendor representatives to assist in item selection.
- Survey the area to determine which materials are most readily available locally.
- Require procurement specialists to publish bulletins on a regular basis, identifying materials and items in short supply on the world market and approximating order-ship-delivery lead times of all equipment and materials regularly used in the contractor work.
- Consider using pre-engineered structures in lieu of specialty designed structures.

ENGINEERING

Choose Configurations That Facilitate or Simplify Handling an Erection

- Require design engineers to develop recommended construction methods and include them with the design to force them to think constructability.
- When designing steel members and connections, remember that erection is much easier if a member to be connected to another can be temporarily positioned on top of the in-place member or on a preinstalled seat on that member before bolting or welding.
- Take advantage of modularization. Vendor-assembled modules are produced under more favorable conditions than those in the field. This will ensure better quality while reducing field erection time.
- Use designs that employ precast concrete components that can be cast in a controlled environment, delivered to the project when needed without intermediate handling, and directly installed.
- Avoid components that require special care and handling in the field.
- Create designs that do not require special care and handling in the field.
- Include special foundations in the design of structures for mounting climbing cranes and elevators if such equipment will be used during construction.
- Locate heavy or bulky items within structures so that as many as possible can be hoisted from a single location of the lifting equipment.
- Maximize the use of straight runs for horizontal lifting and support beams, and avoid curves (particularly complex curves) and angles if possible.
- Consider limitations of standard transport and lifting equipment when designing components. If necessary, design oversized items so they can be fabricated, transported, and erected in parts.

- Use designs that minimize the need for temporary structures such as forming, shoring, bracing, and tie-downs.
- For multiple electrical and piping systems, consider using common utility tunnels or conduits through which multiple systems can be installed (and easily removed or expanded later if necessary) rather than using direct embedment or multiple conduits.
- For multiple foundations in the same area, establish the same bottom elevation for all foundations so that excavation can be handled on a mass basis rather than individually.
- Design engineered items so that they can be dressed out on the ground for installation. In other words, design any components that cross several items (such as ladders or raceways) so portions of them can be preassembled with the engineered item to create a module.
- Design electrical/instrumentation connections with plug-in configurations rather than a labor-intensive connection.
- For complex wiring networks, specify the use of wiring harnesses that are factory assembled and coded.
- In lieu of cast-in-place reinforced walls, consider using the lift-slab technique.
- When designing connections for hydraulic or other systems, create unique designs for each category of connection to avoid any potential for connection mix-ups in the field.
- When designing or specifying large components (such as vessels or rotating equipment), include lifting hooks or other handling devices or features in the design so field erectors will not have to improvise the rigging and handling.
- Provide designs for special measuring devices, templates, or other erection aids that may be useful for aligning or achieving tolerances.

ENGINEERING

Design for Accessibility and Adequate Space

Create designs that promote accessibility and provide adequate space for construction personnel, material, and equipment by doing the following:

- Consider interstitial designs for buildings. This means providing space above all operating floors that is zoned for various operating systems. The vertical clearance in this space should be enough to allow for easy movement of workers. This design greatly simplifies construction and facilities future maintenance and upgrading.
- Locate electrical pull boxes with adequate space around them to simplify cable pulling.
- Size pipe racks to allow easy addition of new lines.
- Incorporate access openings in both exterior and interior walls.
- Provide for reasonable working space around all installed components.

Adapt Designs and Strategies to Project Location and Time

- In an area with a very short construction season or limited labor availability, make maximum use of factory assembled modules and components that have been designed for rapid assembly.
- Consider local labor and specialty contracting capabilities.
- Select designs that best use these capabilities since they will be less costly than imported capabilities.
- If the local population lacks needed skills, maximize the use of remote, off-site fabrication.
- In a union environment, consider jurisdictional rules and wage scales when selecting a design approach.

- Avoid designs whose construction is particularly weather sensitive.
- Avoid the use of materials expected to be in short supply or subject to unusual price inflation during the duration of the project.

Use Realistic Specifications

- Do not require unnecessarily tight tolerances. For example, the imposition of ASTM or nuclear-quality specifications on ordinary construction can be overkill.
- Do not specify an expensive, hard-to-install material when another is far more economical. For example, PVC conduit is lighter, more flexible, and easier to work with than rigid conduit.
- Designers must learn to challenge each specification. Is it the best for the project at hand? A more reasonable specification may be adequate.
- Maximize the use of performance rather than proprietary or descriptive specifications to give greater feasibility to the field.
- Minimize the number of specifications applying to the same type of work, such as concrete, bolt sizes, etc.
- Consider field installation costs in the economic evaluation of material or equipment choices.
- Include in the specification file information on where and why a given specification is applicable. This will assist engineers in selecting the best specification for the job at hand.
- Maintain and continually update a file of lessons learned from previous projects. Make these the subject of training sessions.
- When possible, allow for alternatives in case the primary method or item is not achievable.
- Include requirements for packing and shipping critical items that assure undamaged delivery of them.

- Specify testing methods and procedures that are reasonable for the field.

Assure Quality and Completeness of Design Deliverables (Drawings and Specifications)

- Be willing to hire outside expertise when the in-house staff does not have the talent or time required to prepare quality deliverables.
- Establish a complete system of reviews and checks to ensure accuracy of dimensions, compatibility of drawings and specifications, and consistency of flow diagrams, piping and instrumentation diagrams, etc.
- Use physical or computer models to be sure there are no interference between systems.

Incorporate Safety in Designs

- Specify locations where beams and columns should be drilled to accommodate safety cables.
- Design components to facilitate preassembly on the ground and lifting into final position in modular form.

Construction Phase

Plan and develop the site to promote worker efficiency:

- Use cardboard cutouts that have been cut to scale to represent temporary construction facilities on an overall map of the site drawn to the same scale and brainstorm the best layout of the site to support construction.
- Provide for dust control on roads.
- Develop and stabilize all heavily used foot traffic areas around the construction site.
- Design the construction road network to isolate administrative traffic from traffic that directly supports construction activity.
- If space permits, develop a perimeter road around the site to help prevent traffic congestion and interference.
- Design lay-down areas as a series of alternating roads and narrow lay-down pads that allow any item in the lay-down area to be handled using lifting equipment on the adjacent road
- Shape all lay-down areas for drainage, and construct a supporting drainage network. Stabilize surfaces where material will be placed and spray them with weed killer or cover them with plastic sheeting to prevent grass and weed growth. Make cribbing available for off-ground placement of materials.
- Do not allow long-term storage of any materials adjacent to a facility under construction. Leave clear space around its perimeter that is available for construction equipment and pre-positioning of materials needed for current work activity.
- Locate smoke and dust-producing activities downwind from the center of construction activity.
- Locate/relocate portable facilities to minimize travel distances from worker concentrations.

ENGINEERING

- Regularly clean up and remove construction debris and garbage from work areas.
- Establish grids for construction electrical, gas, water, and compressed air service with distribution points in convenient locations. Design connection trees that are modular and can be moved from distribution point to distribution point as needed.
- Have portable lighting sets available to illuminate work areas during nondaylight hours or where natural illumination is poor.

Perform work when and where it is most efficiently accomplished:

- During construction of multilevel structures, pre-position installed equipment and other materials on the various levels as decks are completed to avoid later problems of access.
- In a congested area where multiple piping and electrical systems are competing for space, install the heavier, bulky components first, leaving the lighter, more flexible items for last.
- When building construction roads, include nondrainage culverts and ducts where future utility lines are expected to be located so that roads will not be cut up later to accommodate laying these lines.
- Fabricate like components, such as rebar cages, on an assembly line basis.

Minimize unscheduled and unproductive activity:

- Use detailed work package planning and adopt a philosophy of never starting on a work package until personnel, materials, and equipment availability is assured.
- Use separate crews for material pickup and spotting at the point of use to keep the supervisor with the crew.
- Obtain old trailers to use in picking up and positioning materials to be used by crews. Materials can be moved

directly from the trailers to the point of placement, thus eliminating multiple handling.
- To avoid the inevitable productivity degradation associated with rework due to changes, on larger projects consider using a special crew within each craft to handle rework, thereby allowing the primary crew to move on to other first-time work.
- If special equipment, such as heavy cranes, must be rented to support certain phases of the project, concentrate the scheduling of work requiring this equipment into as short a time span as possible.
- For larger projects the heavy lifting cranes will need to be positioned near the center of equipment to be lifted and work outward as the project progresses to the edge of the area due to the distance from the crane to the lift point. Access to a heavy lift near the center of a site may not be possible or practical after the majority of the lifts have been completed and might involve bringing in a heavy crane multiple times during a project to complete all lifts.
- Use bar codes and other codes to identify materials in storage to speed up identification time.
- For critical layouts on major projects, consider using two separate survey crews, each starting from the primary benchmark, to lay out construction. This will minimize the potential for layout errors.
- When storing materials in a lay-down area, store them in order of retrieval. This will minimize damage and loss associated with handling and rehandling.
- Paint distinguishing marks (such as a north arrow or "Top") on components to facilitate their final positioning. This will eliminate lost time due to misplacement.
- Assign lay-down areas by discipline.
- Place tool boxes, tool rooms, parts lockers, etc. on wheels or skids to permit their relocation as work moves. Install lifting hooks on them so they can be handled with cranes.

ENGINEERING

- Use bar coding and computerized inventory control to speed tool issuance. This is even more effective if employee ID badges have a bar code so that the employees accessing the tools can be quickly identified.
- Consider using just-in-time material deliveries from suppliers to eliminate the cost and effort of intermediate storage and handling.
- Employ work-saving tools/equipment and modern construction techniques.
- It is impossible to keep up with the market, since new and better technology is always introduced. Ask vendors to demonstrate their equipment. They are usually receptive to providing training as well.
- Use automatic welding machines, nail guns, cordless tools, laser levels, craftsmen stilts, etc.
- Use commercially available material items designed to speed the construction process. For example, commercial forms are available for concrete work, and a complete family of high chairs, clips, and other gadgets can speed the placement and tying of reinforcing steel. Comparable items are available for carpentry, electrical, and other work.
- Have a representative of the organization attend trade shows to learn what is on the market. Bring back literature, and make it available to those with a need to know. Consider making a video tape using scenes from a trade fair or pictures from brochures with appropriate narrative, and distribute this video among the staff.
- Subscribe to trade publications, which contain many advertisements describing innovative products. Prepare a scrapbook of those with the most potential and make it available to the staff.
- Cut out articles from trade and other publications that describe innovative techniques used by competitors. Compile them in a scrapbook that is available to the staff.

Sequence work for optimum efficiency. When the facility to be built includes repeated designs, try to schedule work on repeated elements in series to take advantage of the learning curve:

- Pre-position and temporarily lash heavy or bulky components within a structure when access is most favorable.
- On large concrete slabs, pour sections in checkerboard fashion to reduce the need for forms.
- Install stairs and platforms early so they can be used in lieu of scaffolding and elevators.
- Schedule construction activity around the weather. For example, some buildings may be erected early to provide protected work space for later construction.
- With owner concurrence, construct selected permanent facilities early, and use them for construction support.

Employ construction practices that emphasize safety:

- Erect stair towers early so they may be used for access during construction.
- Use remotely activated release devices on rigging equipment so workers will not have to be hoisted to release them manually from equipment lifted into place.
- Have safety equipment vendors demonstrate available state-of-art safety equipment and provide any training needed.
- Install safety lines and other safety devices on structural members before they are lifted into position.

ENGINEERING

Civil/Structural Constructability Items to Consider

- Be aware of existing site conditions.
- Provide clearing and grubbing of the site.
- Provide drainage and construction roads.
- Provide construction facilities and site access.
- Provide material receiving and lay-down yards.
- Consider overhead crane setup and access.
- Consider modular design and construction.
- Aid the sequence of construction based on design packages.
- Consider repetitive design and construction practices.
- Provide for stair and ladder access to construction platforms.
- Include safety cables and support points for construction.
- Be aware of weather considerations for timing of construction.
- Group design packages to match construction sequence and associated crafts.
- Consider using precast or pre-engineered components.
- Consider anchor bolt templates.
- Provide lifting point loads in the structural analysis.
- Provide monorails and jib-cranes as required.
- Consider maintenance and/or removal of equipment.
- Standardize on anchor bolts, steel members, pipe supports, etc. if reasonable.
- Standardize long pipe racks, provide for expansion, and consider modularization.
- Be aware of existing road crossing clearances and design for appropriate clearances to facilitate transportation of equipment to the new construction site.
- Compare installing permanent roads for construction or temporary roads that will be rebuilt after construction of the new site.
- Provide reasonable specifications.
- Consider using approved "or-equal" materials.
- If a warehouse building is a part of the project, construct early and use as material storage for the project.

- Schedule packages in order to use heavy lift cranes during their window of availability.
- Be willing to change your design if it provides for easier installation, lowers the cost, improves the schedule, and enhances the client expectations as long as it does not compromise the structural capacity or safety of the design.

Notes:

- Not all suggestions and guidelines are available for all situations and locations, especially smaller projects.
- Not all items listed are appropriate for all types of construction projects.
- The above-contained lists and suggestions are for consideration, and those that apply should be used if they enhance the constructability of a project and make it more efficient, reduce the cost and time required, and provide for a safer overall job.

ABOUT THE AUTHOR

The author is a retired engineer with an associate of arts degree in pre-engineering received in May 1969 from East Central Junior College. He continued his education with a bachelor of science degree received in May 1971 and a master of science degree received in August 1975, both in civil engineering majoring in structures from Mississippi State University. He has spent his career working for various industries including chemical, construction, environmental, manufacturing, nuclear, petro-chemical, petroleum, pharmaceutical, power generation, process safety, pulp and paper, railroad, solid waste, steel, and wastewater treatment.

He has worked in all seven USA EPA regions and designed structures and foundations for projects around the world. He has provided numerous consulting services for various clients and has worked closely with the Army Corps of Engineers.

During his career, he was registered as a professional engineer in seven states and currently remains registered in three states. He is a lifetime member of the American Society of Civil Engineers and is a member of the American Institute of Steel Construction. He served on the ASCE task subcommittee that produced "Wind Loads for Petrochemical and Other Industrial Facilities," which was published in 2011. Currently retired, he resides with his wife of fifty years in Baton Rouge, Louisiana.

Printed in the USA
CPSIA information can be obtained
at www.ICGtesting.com
LVHW080557031023
759759LV00050B/476